ACKNOWLEDGMENTS

It is only appropriate that I first acknowledge the guidance and help of my advisor and friend Dr. Jose Principe. Without his support this work would never have been completed. I would also like to thank the members of my committee for their efforts and time spent on my behalf, as well as the members of the Computational NeuroEngineering Laboratory (CNEL).

I must also acknowledge my wife Tammy who was incredibly patient and never wavered in her support of this endeavor. I would also like to thank my children Erin and Matthew who are just too fun to ignore. Although they extended the amount of time required to graduate, I would not trade the time I spent with them for anything in the world. Lastly I should thank my family and friends for not treating me like a dead-beat Ph.D. student.

TABLE OF CONTENTS

Abstract of Dissertation Presented to the Graduate School
of the University of Florida in Partial Fulfillment of the
Requirements for the Degree of Doctor of Philosophy

TEMPORAL SELF-ORGANIZATION FOR NEURAL NETWORKS

By

Neil R. Euliano II

August, 1998

Chairman: Dr. Jose C. Principe
Major Department: Electrical and Computer Engineering

The field of artificial neural networks (ANNs) has reached a point where they are now being used in everyday products. ANNs, however, have been largely unsuccessful at processing signals that evolve over time. Temporal patterns have traditionally provided the most challenging problems for scientists and engineers and include language skills, vision skills, locomotion skills, process control, time series prediction, and many others.

The fundamental concept presented in this dissertation is the formation of temporally organized neighborhoods in ANNs. This temporal self-organization enables the networks to process temporal patterns in a more organized and efficient manner. The concept is biologically inspired and uses activity diffusion to organize the processing elements of the network in an unsupervised manner.

The self-organization in space and time created by my methodology has been applied to three distinct ANN architectures. The new network architectures created by adding the temporal organization are easy to implement and contain properties that are unique in the neural network field. A self-organizing map (SOM) network obtains a unique combination of long-term and short-term memory and becomes organized such that temporal patterns in the input fire sequentially ordered output PEs. These features are utilized in two different applications, a robotic landmark recognition problem and a temporally ordered vector quantization of phonemes in spoken words.

When applied to the neural gas algorithm, the resulting network becomes a dynamic vector quantization network. The network anticipates the future inputs and adjusts the size of the Voronoi regions dynamically. It was used to vector quantize speech data for a digit recognition problem and to predict a chaotic signal. Lastly, the temporal organization was applied to the training of fully recurrent neural networks. It reduces the computational complexity of the training algorithm from $O(N^4)$ operations to only $O(N^2)$ operations and maintains nearly all of the power of the RTRL algorithm. This training method was tested on two inverse modeling tasks and provided a dramatic improvement in training times over the RTRL algorithm.

CHAPTER 1
INTRODUCTION AND PROBLEM DESCRIPTION

This dissertation focuses on neural network architectures and training methods for processing signals that evolve over time. The fundamental concept underlying the techniques described herein involves the formation of neighborhoods where temporally correlated processing elements (PEs) are clustered together. We have applied this concept to three different neural network architectures and found that it improves the performance of each one – either by increasing the functionality of the neural network or by improving its training.

This chapter contains a description of the problem as well as background information that will help describe the shortcomings of the present methods. Chapter 2 presents a review of the relevant literature necessary to understand the material in the context of the current state of the art. Chapter 3 contains the theoretical description of the techniques and networks proposed by this work, including a few simple examples to elucidate the fundamental concepts. In Chapter 4, six more extensive and practical problems are solved using the temporal neighborhood concepts. These examples include speech recognition, chaotic prediction, system identification and control, and robotics. Chapter 5 concludes the dissertation with a summary of the work and possible future research directions.

Temporal Processing

Most scientific problems can be grouped into two domains, static and dynamic problems. Static problems consist of information that is independent of time. For instance, in static image recognition, the image does not change over time. On the other hand, time is fundamental to the dynamic problem. The output of a dynamical system, for example, depends not only on the present input but also on the current state of the system, which encapsulates the past of the input. Temporal processing is the analysis, modeling, prediction, and/or classification of systems that vary with time. Patterns that evolve over time have traditionally provided the most challenging problems for scientists and engineers. Language skills (speech recognition, speech synthesis, sound identification, etc.), vision skills (motion detection, target tracking, object recognition, etc.), locomotion skills (synchronized movement, robotics, mapping, navigation, etc.), process control (both human and mechanical), time series prediction, and many other applications all require temporal pattern processing. In fact, the ability to properly recognize or generate temporal patterns is fundamental to human intelligence.

Traditional analysis models (ARMA models, etc.) are well known but are usually linear and require significant expertise on the subject and a strict correspondence between the studied process and the constructed model. Artificial Neural Networks (ANNs) offer robust, model-free methods without requiring as much application specific expertise. Secondly, neural nets are adaptive (similar to ARMA models). This is a natural way to compensate for the drift of measuring devices and slow parameter changes inherent in real systems. Thirdly, neural nets are naturally parallel systems that offer more speed in computation and fault tolerance than traditional computing models. [Kan94]

Most of the major neural network success, however, has been mainly in the realm of static, instantaneous mappings (for example, static image recognition or pattern matching). Conventional neural net architectures and algorithms are not well suited for patterns that vary over time. Typically, in static pattern recognition a collection of features – visual, semantic, or otherwise – is presented and the network must categorize the input feature pattern into one or more classes. In such tasks, the network is presented with all relevant information simultaneously. In contrast, temporal pattern recognition involves processing patterns that evolve over time. The appropriate response at a particular point in time depends not only on the current input, but also potentially on an unspecified number of previous inputs.

Static ANNs have been modified in various ways to process time-varying patterns, typically by adding short-term memory to the static pattern classification ability of the various architectures. The short-term memory holds onto some of the past events so that the static ANN can then classify or predict the temporal pattern. As I will explain in the next few sections, however, these hybrid structures (memory added to static architectures) have not been widely successful in the various temporal processing areas.

Static Supervised and Unsupervised Learning

The purpose of neural processing is to capture the information from an external signal in the neural network structure. This is a form of organization. It can be accomplished in an unsupervised manner using only the input, or in a supervised manner guided by an extra input called the desired signal. Unsupervised training can only extract information from the input signal whereas supervised training can learn mappings between the input signal and the desired signal. They differ in the methods, but at the core they share the same function, learning a representation of the external world.

The most common supervised network is the multilayer perceptron (MLP) which uses the error back-propagation [Rum86] learning algorithm. The MLP is characterized by layers (input, hidden, and output) of processing elements (PEs) that have a smooth non-linearity at their output. The nonlinear output of the MLP PEs is what differentiates the MLP from a typical adaptive filter. It provides the capability to map problems that are not linearly separable. In fact, it has been proven that an MLP with one hidden layer can uniformly approximate any continuous function with support in a unit hypercube [Cyb89]. Like in adaptive signal processing using the LMS algorithm, the back-propagation algorithm applies a correction $\Delta w_{ji}(n)$ to the synaptic weight $w_{ji}(n)$ that is proportional to the gradient of the error $\partial \xi(n) / \partial w_{ji}(n)$. The chain rule is used to recursively calculate the error for each layer of the network.

Unsupervised networks are typically based on or derived from Hebbian learning. Hebbian learning is a biologically inspired learning rule that finds the correlations present in the input data. Because unsupervised networks can extract information only from the input, they are typically used for data analysis and preprocessing. They cannot reliably be used directly for classification since a labeling of the inputs is required for classification. Both supervised and unsupervised learning will be described in detail in Chapter 2.

Adding Memory to Neural Networks

How do you use a static neural network architecture to process temporal patterns? The answer is to simply add memory. Without an appropriate memory to store information from the past, a neural network is limited to static pattern recognition or function approximation. The key questions that need to be answered while creating temporal neural networks are what type of memory do you use and how is the memory integrated into the training algorithm.

Memory in neural networks can be classified into two categories: short-term memory and long-term memory. Short-term memory typically involves a representation of the temporal data, usually by creating multiple copies of the input data at various time delays (e.g. tapped delay line). Long-term memory, on the other hand, is the storage of information from the past into the structure of the network. For example, over time, the training of the network captures information about the input signal and this information can be considered long-term memory. Another example of long-term memory is the storage of patterns in an associative memory. Long-term memory corresponds more closely with the traditional biological concepts of memory. The main difference between the two is that the short-term memory is used for signal representation while the long-term memory is a trained memory that typically cannot represent unknown patterns. Another way to differentiate the two is that short-term memory is usually described by activations of nodes or taps (dynamical information), and long-term memory is stored in the weights of the network (statistical information).

Most of the work in temporal ANN research has focused on the application of short-term memories since they provide a mechanism to represent a temporal pattern in a static manner. For instance, a tapped delay line converts a temporal signal into a static pattern (the present input and the N past inputs) which can then be processed by a standard static ANN. Most short-term memory techniques fall into two categories. The first is to explicitly add memory structures and the second is to use recurrent loops in the network to save information. Long-term memory (the weights) has largely been ignored by the ANN research community for the storage of temporal patterns, but I will use it to store temporal correlations in the structure of the network.

Short-Term Memory Structures

The simplest form of memory is a buffer containing the *N* most recent inputs. This is often called a *tapped delay line* or a *delay space* embedding and forms the basis of traditional statistical autoregressive (AR) models, as well as dynamical system state space manipulations. This is a very popular model and has been used in many applications. The time-delay neural network (TDNN) [Wai90] uses a tapped delay line to convert the temporal pattern into a spatial pattern allowing the architecture to be trained using only standard back-propagation methods. The TDNN, however, has several drawbacks. First, the length of the delay line must be chosen a priori, we cannot work with arbitrary length sequences. In addition, the TDNN requires that the data is properly registered in time with the clock controlling the shift register. It imposes a rigid limit on the duration of patterns and suggests that all input vectors be the same length. Most importantly, two patterns which are very similar temporally (e.g. shifted one step in time) will be very different spatially, which is the metric used by ANNs. For example, [1 0 0], [0 1 0], [0 0 1] are temporally shifted but are spatially on the corners of a unit cube.

Using decay traces or exponential kernels to sample the history of the input helps alleviate some of the problems with the TDNN. A common methodology to describe the various memory architectures is to represent the short-term memory as a convolution of the input sequence with a kernel function, k_i: $\bar{x}_i(t) = \sum_{\tau=1}^{t} k_i(t-\tau)x(\tau)$, where x(t) is the input. Tank and Hopfield [Tan87] proposed a set of Gaussian kernels that are distributed over time with varying means and widths to sample the time history. The gamma model [DeV91] is an example of an exponential trace memory that uses the set of gamma kernels. The exponential trace memory has a more smooth representation of the past of the input since it decays exponentially. It gives more strength to the more recent inputs. The gamma memory also has a tunable parameter that trades off *depth* for *resolution* when the system requires information from farther in the past. Depth roughly refers to how far back into the past the memory stores information and resolution refers to the degree to which information concerning the individual elements of the input sequence are preserved. The exponential trace memories can be computed incrementally and easily, thus greatly increasing its usability. Viewing memory in this way, as a kernel function passed over the input, one can see that almost any kernel function will result in a distinct form of memory.

The main problem with all of these memory architectures, however, is that they are all "prewired" one-dimensional cascades of delay elements. TDNNs are also known to train very slowly.

Theoretically, memory added to a system can be thought of as creating an embedding of the dynamics into a space larger than the original input space. An embedding of a dynamical system is based on the similarity between delays and derivatives (the first order approximation to a derivative is the difference between the signal and the delayed signal). The delayed values of a single variable can be used to represent the dynamics of a multi-dimensional system. Conceptually this can be rationalized as combining the first-order differential equations for the system (state space description) into a single high-order differential equation for one variable and then using the delay technique to approximate the derivatives of this equation – giving a new representation of the system states. This mathematical construct is effective but not necessarily efficient. For example, a dynamical system requires a minimum of $2D+1$ taps to preserve the dynamics of a D dimensional system [Tak81]. If the dimension of the system is unknown, as is often the case, a large embedding is usually used. The embedding also does not efficiently encode the input ordering. It does a time-to-space mapping that treats the temporal information the same as a spatial input, allowing for all permutations of the order of inputs without regard to the limitations imposed by the dynamics of the system. The gamma memory and other convolution memory kernels warp or rotate the embedding space to more accurately (or efficiently) represent the system dynamics. A proper use of the embedding methodology requires a significant amount of work to determine a number of parameters, including the number of taps, the time between taps, the time between vectors, and the number of data samples. This is rarely done.

Recurrent Networks

The MLP and TDNN are both feedforward networks where the data flow in the network moves strictly forward. No feedback is used. The feedback in recurrent networks can also create memory. The important distinction between the two types of memory is that memory created with feedback can be adapted and trained on-line, creating a flexible and adjustable memory mechanism. Feeding back outputs between different layers can lead to a generalization of storing not only the input but the "state" of the network (i.e. a processed version of the input) [Elm90][Moz94]. In theory, the recurrent architecture is sufficiently powerful to handle arbitrarily complex temporal problems. The focused memory architectures such as the TDNN can also [San97], but may require a very large number of taps and weights.

In practice, however, recurrent networks are much more difficult to train than the static networks. The recurrency adds tremendous power to the network (any memory architecture can be created with a recurrent neural network). This power, however, creates very complicated error surfaces. In recurrent networks, the states of the PEs in the network affect both the output and gradients. Therefore calculating the gradients and updating the weights of a recurrent network is a much more difficult and time consuming process.

Because of these difficulties, the mainstream engineering community has largely ignored recurrent networks. Recently, however, the recurrent networks are being used more and more as engineers reach the limits of the capabilities of TDNNs and other simpler architectures. Recurrent networks are hot topics in the fields of dynamic modeling and control.

Training Dynamic Neural Networks

Recurrent networks, either fully recurrent or partially recurrent (e.g. the gamma network), cannot directly use static backpropagation methods since the time history of the network and its inputs are critical to the outputs produced by it. Static backpropagation computes only the gradients based upon the current inputs and outputs. To train a dynamical system, the past information is at least as important as the present and thus a temporal backpropagation technique must be used. Recurrent backpropagation (fixed-point learning) can be used to train a general recurrent network to move to stationary states. Its assumption of constant inputs and an approach to an attractor, however, precludes the recurrent back-propagation algorithm from real-time temporal processing.

The TDNN can use static backpropagation because its memory is fixed and is at beginning of the network. The tapped delay line can be thought of as a temporal preprocessor converting dynamic patterns to static patterns, thus the network is trained in a completely static manner. Most other temporal networks, however, are trained using one of two first-order temporal methods: back-propagation through time (BPTT) [Rum86] or real-time recurrent learning (RTRL) [Wil89]. Both of these methods are gradient descent methods. The RTRL method brings the activations and their derivatives forward in time until the desired signal is available, and the BPTT method propagates back the errors from the desired signal to the beginning of the pattern. RTRL recurrently calculates the gradients of each PE with respect to every weight. This process allows on-

11

line updates (updates every sample), but calculating all the gradients is a time consuming process. In fact, if there are N fully recurrent PEs in a network, the RTRL algorithm requires $O(N^4)$ operations per sample. The BPTT method requires fewer computations, but is non-causal. Thus it cannot be directly implemented in an on-line fashion.

Both methods suffer from the following problems:

- The computation of the gradient must occur over time. But the nonlinearity in each layer (actually it is the derivative of the nonlinearity required for the gradients) attenuates these gradients. Thus, if information is required from more than a few samples in the past, these training methods may have a difficult time maintaining and using this information. As the errors are propagated, the gradients get small and the impact of a connection weight – even if appropriate—will be masked by other weights if their values are inappropriate. This is true for large feedforward nets as well, but the feedback nature of the recurrent network in time makes this a much bigger problem in recurrent networks.

- The desired signal must be defined over time. For example, how do you define a target signal when trying to detect a sequence? If the target is high throughout the pattern, the network will recognize partial sequences. But if the target is high only at the end, the network may be punished for partially recognizing a major portion of the sequence.

- Temporal backpropagation is inherently slow both computationally and in terms of the number of training samples required to find an adequate solution.

Recently, second order gradient methods like the recurrent least squares (RLS) and the extended Kalman filter have been used in order to reduce the number of training samples required for a good solution. These methods use second order gradient information to determine more accurate data on the shape of the performance surface at the current operating point. This allows for much faster convergence but requires more computations per sample. These second order gradient methods still need to compute the dynamic gradient information and thus suffer from the same problems listed above.

Summary of Problems with Standard ANN Architectures

In summary, the standard ANN architectures when applied to temporal processing suffer from problems with supervision and problems with short-term memory. The problems that can be attributed to supervised training include:

- The problem of assigning credit or blame to actions when the overall success or failure of the system results from a series of actions and cannot be judged instantaneously (i.e. how do you design a target signal?).

- Back-propagation training can be very slow, often requiring thousands of training epochs. This problem is derived from many sources. The backpropagation algorithm must either take small steps in the gradient descent or use more computationally intensive error calculations (higher order derivatives). Since all nodes in a network are typically learning independently, several problems may occur. First all the hidden nodes may move together to try to solve the largest source of error, instead of dividing up the problem and each solving a different portion. Second, once the nodes have divided the problem, each tries to solve their portions independently. The movement of each node through the error surface affects all the other nodes, creating a moving target for each node. Thus instead of a direct movement of the nodes to useful roles, we see a "complex dance" among all units [Fah91].

- Recurrent back-propagation trains even slower for many reasons. First, the training methods require more computations than the static backpropagation. Second, the error gradients tend to vanish exponentially as they are propagated through time. Thirdly, the recurrent networks tend to have a much more complicated error performance surface with many local minima, making the gradient search very difficult.

- Supervised techniques require presegmented and prelabeled training data. This often must be done by hand and is quite time consuming. The rule of thumb for ANN training is 10 training exemplars for each adjustable weight. Thus for large networks, finding enough training data is a difficult task. If there is an insufficient amount of training data, the network will tend to memorize the data rather than draw reasonable generalizations about the data.

Problems related to short-term memory structures include the following:

- The common short-term memory techniques (tap delay lines, etc.) use a time-to-space mapping to represent the past of the signal. By converting time into just another spatial dimension, the unique features of the temporal information are lost (e.g. continuity, limitations based on the dynamics of the system, etc.). The short-term memory is a representation of the data, not a memory structure.

- The typical short-term memory structure is a rigid architecture that must be pre-wired.

- Short-term memory structures typically add many weights to the input (or interior) layer (e.g. A TDNN with N taps will create N times more weights in the first layer), which exacerbates the problems with the performance surface and the amount of training data. The resulting networks tend to have so many degrees of freedom that they do not generalize well (i.e. memorization due to insufficient training exemplars).

The Approach

It is a Herculean challenge to attempt to solve all of the above problems. This work focuses on a method of self-organizing PEs in a network architecture based on their temporal correlations. This concept is biologically inspired and has been applied to three different types of neural networks. By creating temporal neighborhoods of PEs in the architecture, we have increased the performance of the networks – either through increased functionality and power or through better training methods.

When this technique is applied to a self-organizing feature map (SOFM or SOM), the temporal neighborhoods create traveling waves of activity which diffuse through the PEs. The resulting architecture has a spatio-temporal memory that is selective and recognizes temporal patterns similar to those it has been trained with. The typical ANN memory simply embeds the data for further processing by the ANN, without any mechanism for recognition. This architecture, however, is similar to biological memories in that it responds preferentially to known temporal patterns – this is unique in the neural network literature.

When the temporal neighborhood approach is applied to the neural gas algorithm, the network becomes a temporal vector quantizer that again responds preferentially to known temporal patterns. The temporal vector quantizer uses the past of the signal to anticipate the next input by expanding the Voronoi region associated with the expected next input. This allows the network to remove noise in the signal and generate better vector quantization based upon the temporal training and recent past of the signal. This anticipation is similar to how the human brain deals with noise in its environment – it uses the past to predict the future and correlates what it is sensing with this prediction. This is part of the reason humans can understand speech in very noisy environments, and also why two people can hear completely different things from the same set of sounds.

When we apply the technique to the training of recurrent neural networks, the new training technique reduces the computational complexity of the RTRL algorithm from $O(N^4)$ to $O(N^2)$. This dramatic improvement comes with only a slight increase in the number of iterations of training data required. The overall speed-up taking into account both the decreased computational complexity and increased number of training samples is still dramatically better. In fact, the $O(N^4)$ property of the RTRL algorithm makes it virtually unusable for sizeable networks.

In general, the self-organizing nature of the temporal neighborhoods helps alleviate many of the problems with the supervised techniques. Additionally, the novel spatio-temporal memory architectures provide a unique methodology for solving the problems with short-term memory.

CHAPTER 2
LITERATURE REVIEW

This chapter presents background information and a literature review of topics

that either influenced this work, relate to this work, or will be compared and contrasted

with this work. The chapter begins with a presentation of current research on biological

neural networks and methods of temporal processing. This section is important because it

motivated my work. I do not, however, claim that my work is biologically feasible or

occurs in nature. Next, this chapter contains a description of the state of temporal neural

network research. Since most of the work in temporal neural networks takes the form of

extensions to static neural networks, an overview of static neural network learning is also

presented. The contrast between the biological and artificial neural networks and the way

they process time is striking. Static artificial neural networks are very similar to the static

characteristics of real neurons, but temporal neural networks share little in common with

their biological counterparts.

Biological Research

This section contains a description of biological neurons and their temporal

characteristics, as well as other biological mechanisms that may help in processing time

based signals. Recently, there has been extensive research into the temporal

characteristics of the brain as well as in learning dynamics. This research has not yet

16

been integrated into the artificial neural network community, but holds promise for creating powerful, temporal ANNs. This information provides a motivation for the main principal of this work – that the creation of temporally organized neighborhoods in a neural network improves the performance of the network for temporal processing. The concept of diffusing temporal information through the network is one of the fundamental concepts used to rationalize the formation of these neighborhoods.

<u>Neurons and Learning</u>

Fundamentally, the artificial neural network is modeled after a collection of neurons in the brain. Each neuron is composed of three basic components: the cell body, the dendrites and the axon. [Fre92] The dendrites are a widely branching set of filaments that collect information from other neurons. The axon is a long transmission medium that contains fewer branches and transmits the output of the neuron to other neurons. Synapses are the junctions between axons and dendrites. The dendrites collect incoming pulses from other synapses, convert them to currents and sum them all at the initial segment of the axon. This summation works across both dendritic space (summation over all the dendrites) and across time. Each synaptic membrane acts as a leaky integrator with an associated time constant. The critical function of the axon is to transmit the time-varying amplitude of current summed by the dendrites to distant targets without attenuation. [Fre92] If the neuron reaches a certain threshold, it *fires* or *depolarizes*, which means that it produces an energy spike on its axon. The firing contains a refractory period such that a constantly active neuron will produce an impulse train on its axon. How biological neural networks are trained is not well known, but most of what is known about the training is based on the Hebbian learning concept (which will be discussed later). The Hebbian learning law strengthens synapses (allowing more responsiveness from the post-synaptic neuron) when the two neurons fire at the same time. If there is a consistent correlation between the firing of two neurons, then the pre-synaptic neuron must be at least partially responsible for the firing of the post-synaptic neuron.

A static artificial neural network is modeled loosely on an interconnected cluster of neurons. Each neuron is modeled by a processing element (PE) and a set of connections between processing elements. Typically, a processing element simply sums the inputs, nonlinearly warps the output, and then passes this output to its downstream connections. Training is implemented in either an unsupervised manner, usually using a form of Hebbian learning, or in a supervised manner, which has no biological parallel. Notice that *none* of the temporal characteristics of a neuron are used in static neural networks or their temporal extensions.

Recently, there has been significant work on a more complete modeling of individual neurons and their temporal characteristics. Christodoulou and others [Chr95a][Chri93] have modeled the biological neuron including the random spiking nature, excitatory/inhibitory synapses, the transmission delay down the axon, and especially the membrane time constant. The membrane time constant is the main temporal property modeled today. Most modeling approaches use simplifications of the Hodgkin-Huxley equations that result in a leaky integrator model of the neuron membrane potential. This is an important feature of biological neurons, since the past history of the signal remains active on neurons for a short period and can influence the result of future inputs.

Additionally, the gas nitric oxide (NO) has been found to be involved in many processes in the central nervous system. One such process is the modification of synaptic strength thought to be the mechanism for learning (and most commonly used in ANNs). Neurons produce NO post-synaptically after depolarization. The NO diffuses rapidly (3.3 x 10^{-5} cm^2/s) and has a long half-life (~4-6 seconds), creating an effective range of at least 150 μm. Large quantities of NO at an active synapse strengthen the synapse (called *Long Term Potentiation*, or LTP). If the NO level is low, the synaptic strength is decreased (*Long Term Depression* or LTD) even if the site is strongly depolarized. NO is thus commonly called a *diffusing messenger* as it has the ability to carry information through diffusion, without any direct electrical contact (synapses) over much larger distances than normally considered (non-local). The NO diffusion and non-linear synaptic change mechanism has been shown to be capable of supporting the development of topographical maps without the need for a Mexican Hat lateral interaction (described later). This seems to be a more biologically plausible explanation of the short range excitation and long range inhibition than the preprogrammed weights of synaptic connections which are typically assumed to implement the same effect [Kre96a][Kre96b].

In addition to the possibility of lateral diffusive messenger effects, the long life of NO can produce interesting temporal effects. Krekelberg has shown that NO can act as a memory trace in the brain that can allow the temporal correlations in the input to be converted into spatial connection strengths. [Kre96b] This mechanism for capturing the temporal correlations of the input using an NO diffusion process is similar to the method I will present in more detail in Chapter 3.

Hippocampus

The hippocampus is the primary region in the mammalian brain for the study of memory and learning because: [Bur95]

- hippocampal damage causes memory loss,

- the hippocampus is the simplest form of cortex

- long-term potentiation (LTP) has been found in the hippocampus (synaptic plasticity)

- cell firing in the hippocampus is spatially coded (place cells).

- all sensory inputs converge on the hippocampus and the output from the hippocampus is extensively divergent with projections onto most of the cortical areas.

Figure 2-1 shows the major subfields and their projections of the hippocampus. The hippocampus is formed from sheets of cells, with most of the interconnections contained in these sheets (minimal connections between sheets). Most projections have large divergence and convergence, except the dentate gyrus to CA3 projection which has

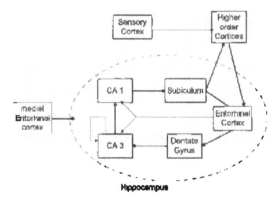

Figure 2-1: The major subfield of the hippocampus

mossy fiber projections from each granule cell, making very large synapses onto only 14 or so pyramidal cells. Hebbian LTP has been observed in much of the hippocampus. A variety of interneurons provide feed-forward and feed-back inhibition.

One of the most interesting (and for this work, most relevant) aspects of the Hippocampus is that it contains "place cells" and other functional clusters of neurons. Place cells are small patches of neurons that selectively fire only when the animal is in a specific location of its environment. These are groups of thousands of neurons that fire together and are linked to other place cells. As the subject moves through a familiar set of locations, the patches fire sequentially and the linking of these patches allows for predictive navigation. They have been found in fields CA3 and CA1 of the rat hippocampus. [Bur93] These place cells are temporally and spatially organized neurons that are correlated in their reaction to temporally occuring events.

Diffusion Equations (Re-Di Equations)

The diffusion equation (or the reaction-diffusion equation if the medium is active) can be used to explain certain characteristics of a neuron and neuronal clusters. In its generic form, however, it is used in many other fields.

Objects such as cells, bacteria, chemicals and animals often have the property that each individual moves about in a random manner (e.g. brownian motion). When a concentration of these objects occurs, this random motion causes the objects to spread out into lower concentration areas of the environment. When this microscopic movement of the group results in macroscopic motion, we call it a diffusion process. If we assume a one-dimensional motion and a random walk process, we can derive the diffusion equation

22

from a probabilistic treatment of the process. By finding the probability $p(m,n)$ that a particle reaches a point m steps away at n time steps in the future, we find the distribution of particles at time n. Using the random walk assumption and allowing n to be large, it can be shown that the resulting distribution is the Gaussian or normal probability distribution:

$$p(m,n) \sim \left[\frac{2}{\pi n}\right]^{1/2} \exp\left[\frac{-m^2}{2n}\right], \quad m \gg 1, \ n \gg 1$$

Next, we determine the probability of finding a particle in an area between (x-Δx, x+Δx) at time t by rewriting the equation for $p(m,n)$ as the sum of the probability of moving right from x-Δx at time t-Δt or moving left from x+Δx at time t-Δt. If we take the partial of p with respect to t and allow $\Delta x \to 0$ and $\Delta t \to 0$ we obtain the diffusion equation:

$$\frac{\partial p}{\partial t} = D \frac{\partial^2 p}{\partial x^2}$$

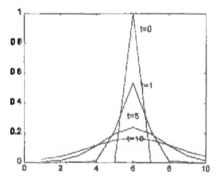

Figure 2-2 - Diffusion process

where D is the diffusion coefficient which defines how fast the particles spread. A typical diffusing process creates a spreading of a concentration into ever shallower and shallower Gaussians as shown in Figure 2-2.

The reaction-diffusion equations were originally proposed by Turing in 1952 and are typically used to explain natural pattern formation [Tur52]. They have been used to model insect populations, the formation of zebra stripes, crystal formation, galaxy formation and many other naturally occurring patterns and self-organizing systems. Turing's proposal modeled patterns found in nature by an interaction of chemicals called "morphogens". The different morphogens react with each other AND diffuse throughout the substance via the equation:

$$\frac{\partial m_i(x,t)}{\partial t} = f(m_i(x,t), m_j(x,t)) + D_{m_i} \frac{\partial^2 m_i(x,t)}{\partial x^2}$$

where $m_i(x,t)$ is the concentration of morphogen i at time t, D_m is the diffusion coefficient, and $f(m_i, m_j)$ is a function (typically nonlinear) that represents the interaction between morphogens. By varying the interaction between chemicals and the speed of diffusion, complicated spatial patterns of chemicals are created.

The reaction-diffusion equations have also been used to explain *traveling waves* such as the traveling impulse down the axon of a neuron. If the reaction portion of the Re-Di equations represents the kinetics of the system and these kinetics are nonlinear, then the system can create a traveling wave. One requirement for a traveling wave is that the kinetics of the system are *excitable*, where excitable implies two stable states where a small excursion away from one state may drive it to the next state. Another requirement is that after excitation, the system must relax back to the original state. An example of such a system is the Fitzhugh-Nagumo equations (FHN) that are a simplified version of the Hodgkin-Huxley model that describes the transmission of energy down the axon of a neuron. The FHN equations can be described by the following system of 3 equations [Mur89]:

$$\frac{\partial u}{\partial t} = f(u) - v + D\frac{\partial^2 u}{\partial x^2}$$

$$\frac{\partial v}{\partial t} = bu - \gamma v$$

$$f(u) = u(a - u)(u - 1)$$

where u is roughly equivalent to the membrane potential, v lumps the effects of most of the ionic membrane currents, and a, b, and γ are constants. The null clines of the kinetics in the (u,v) phase plane are shown in Figure 2-3.

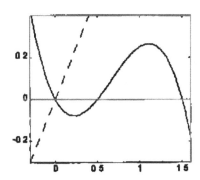

Figure 2-3: Null clines for dynamics of FHN equations

The general concept is that when one element fires, its activity is diffused to its neighbors and pushes them just far enough from their stable state to move them to the "excited" state. Next, these newly excited elements excite their neighbors, etc. The elements which were excited originally then begin to relax, creating a traveling wave of activity. The traveling wave from the FHN equations is shown in Figure 2-4. In this case, not only does the system relax, it also has a refractory phase which inhibits future excitation for a period of time. [Tys88]

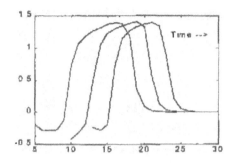

Figure 2-4: Traveling waves caused by the FHN equations

Diffusion and other biologically plausible local communication techniques have increasingly been used in neural networks. For example, the Kohonen algorithm can be implemented in analog hardware with an active medium using diffusion [Ruw93]. Diffusion has also been used frequently in visual imaging systems [Cun94]. Sherstinsky and Picard have proposed a cellular neural network based on Re-Di equations that can solve optimization problems [She94]. On key aspect of this work is that diffusion in the PE space of a neural network allows temporal information to be transmitted and stored using only local communication. This is similar to the diffusion of NO in the brain which is thought to affect the plasticity of synapses in areas where many neurons are firing at once. Without direct connectivity between two PEs, communication and temporal memory can be implemented using the local storage and transmission of a diffusing object (in our case, diffusing activity).

Biological Representations of Time

Another example of neurobiological research that has not been used in ANNs is the concept of rhythm. Recently, there has been some interesting research on oscillators, central pattern generators, rhythm and their effect on human pattern recognition. Rhythm has been studied in biology and found that rhythmic signals from insects can be entrained or phase-locked to an external rythmic pattern – without high-level processing (the patterns are faster than the minimum response latency) [McA94]. There is evidence that the dynamics of many biological systems have natural rhythms that share the same frequency. Communication and locomotion, for instance, are highly dependent on rhythm and pacing. It has also been suggested that EEG rhythms play an important role in learning and temporal recognition. For instance, neurons are thought to modify their synaptic strengths only when the θ rhythm is in the correct phase. The θ rhythm is a sinusoidal component of the EEG that ranges from 7-12 Hz. The θ rhythm has been linked with displacement movements (e.g. walking) and many other repetitive actions. Since the θ rhythm must propagate through the neural tissue, this also could play the role of a moving wavefront that controls learning.

Rhythm can be thought of in two ways: either as an external pacemaker that synchronizes the network in some fashion, or as the output of a collection of neurons that are working in unison. For the first case, there is little if any research on the effects of an external pacemaker on temporal ANNs. The pacemaker would create a time-varying network where the output of the network is dependent on the time or phase of the pacemaker. The pacemaker could also act as a sampling signal. For instance, learning may only occur at a specific phase of the θ rhythm. In the second case, the rhythm could be the result of synchronized processing. For instance, waves of activity in the brain could be caused by the processing of the spatio-temporal patterns constantly input to the network by the continuous motions of the eyes and other sensory muscles.

Stanley and Kilmer [Sta75] have proposed a "wave mode" of memory that can learn sequences. It is based on the anatomy of the dentate gyrus (in the mammalian hippocampus) and can be summarized as follows:

- The hippocampus is organized into transverse slices called lamellae

- The majority of connections in the hippocampus do not leave a lamella (small longitudinal spread)

- Sensory inputs arrive via the perforant path to excite cells directly

- A small number of mossy fibers connect cells longitudinally (across lamellae)

- Cells excited by an input spread excitation to its neighbors, causing a wave of activity to travel down the cell's lamella

The wave formation is based on the pyramid and granule cells receiving excitatory influences from the hippocampal input pathways that in turn excite interneurons whose axons inhibit the pyramid and granule cells. This excitation and inhibition create the waves of activity in the lamella.

The memory is created by the association of the various waves in different lamellae via the mossy fibers that interconnect the lamellae. Each wave is created by a sensory input that triggers a cell in a lamella and can move a short distance before dying. Randomly distributed mossy fibers interconnect the lamella. The connection weights are strengthened in a Hebbian manner – when two waves from different lamella are coincident with a connecting mossy fiber, this connection is strengthened. Thus, the next time the first input wave passes the same position, it can automatically trigger the second wave even without the corresponding input. This is shown in Figure 2-5. For longer temporal relationships, one wave will trigger a second wave in another lamella via pre-

strengthened longitudinal connections which will continue after the first wave has died.

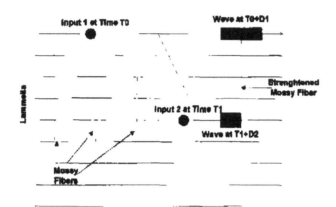

Figure 2-5: Stanley and Kilmer's wave model [Sta75]

Biological Models for Temporal Processing

Living neurons act as leaky integrators with time constants on the order of tens to

hundreds of milliseconds. This can lead to the storage of information in a way that may

lead to temporal sequence processing. Most ANN temporal methods store the

information in a spatial manner. The spatial approach to signal storage is used in the

brain for auditory and visual processing (e.g. SOMs). The way in which these maps are

then processed is not necessarily spatial. Reiss and Taylor propose an interesting

temporal sequence storage mechanism based on a leaky integrator network [Rei91]. The

basic concept is to use the leaky integrator neurons as temporary storage for an

associative memory that is implemented like a single layer neural net. The network has

been shown to have a capacity proportional to the number of neurons. The problem with

this network is that the connection matrix seems to be very heavily skewed to only

predicting the next input with little information from further in the past. This is similar to a simple state machine or markov chain. An interesting part of this work is the possible connection to the function of the hippocampus. The memory network corresponds to the dentate gyrus, the CA3 corresponds to the predictor, and the input line is similar to the perforant path (between EHC, DG, and CA3).

Kargupta and Ray proposed a temporal sequence processor that is based on the reaction-diffusion equations. [Kar94] Drawing an analogy between chemical diffusions in biology and spatio-temporal sequential processing, their model is based on a collection of cells that react to different inputs. When a cell becomes active (by recognizing its input), it outputs its own specific chemical. This chemical diffuses throughout the medium containing the cells. Each cell contains a memory of the chemical makeup at its location when it fires. The background medium thus stores the temporal history of the signal by diffusing all the various chemicals. This approach is more of a chemical model than an information processing model and has several difficulties when applied to realistic problems.

Static Neural Network Learning

This section contains a summary of the static neural network learning mechanisms. Almost all of the work in temporal ANNs is based on the principles from static ANNs. Since unsupervised training is most similar to known biological learning mechanisms, it will be presented first. Unsupervised learning does not have a desired signal and extracts information only from the input of the signal. As such, unsupervised techniques typically do not directly implement classifiers, but are usually used for preprocessing the input. For example, unsupervised networks can be trained to perform principal component analysis (PCA), vector quantization (VQ), and data reduction. Supervised learning is presented next and these algorithms use a desired signal to train the network to mimic the desired input-output map. The desired signal can be thought of as a teacher or external influence that guides the network to the desired state. As we mentioned before, there is no known biological analog to supervised training.

Unsupervised Learning

Most unsupervised (also known as competitive or self-organizing) learning is based on Hebbian learning. *Hebbian learning* is derived from the work of the neuropsychologist Hebb who noted in 1949 that when cell A repeatedly participates in the firing of cell B, a growth process occurs between the two cells which increases the efficiency of the link between cell A and cell B. This can be stated as "neurons that fire together, wire together". This mechanism is often called correlation learning because the links are increased when there is a statistical correlation over time between the presynaptic and postynaptic activities. To avoid excessive weight growth, Hebbian synapses typically also include a decrease in the strength of a connection between two cells which are uncorrelated. Conversely, *anti-hebbian learning* is a learning rule that is based on increasing the strength of a connection when the presynaptic and postynaptic signals are negatively correlated and weakens them otherwise.

A typical expression for Hebbian learning is

$$\Delta w_{k_j}(n) = \eta y_k(n) x_j(n)$$

where w_{k_j} represents the synaptic weight between cell k and cell j, x_j is the presynaptic activity and y_j is the postynaptic activity. η in the above equation is the learning rate.

This rule, however, does not include the weakening of uncorrelated signals, and thus the weights will forever increase. Introducing a nonlinear forgetting factor into the equation can control the weight growth:

$$\Delta w_{kj}(n) = \eta y_k(n)x_j(n) - \alpha y_k(n)w_{kj}(n)$$

where α is the decay constant. This equation can be rewritten as:

$$\Delta w_{kj}(n) = \alpha y_k(n)\left[cx_j(n) - w_{kj}(n)\right]$$

which is the standard Hebbian learning rule. Notice that when the postsynaptic neuron fires, w_{kj} moves toward x_j exponentially. By manipulating the definitions of the variables, this equation can be reformulated into the *competitive learning* rule.

In competitive learning, a group of neurons are clustered such that one and only one neuron wins a competition for each input. Algorithmically, the winner is simply selected by choosing the PE with the highest/lowest output, which can be physically implemented using lateral inhibition between nodes. Biologically, neurons fire in clusters and the competition between clusters is believed to be due to long range inhibition and short range excitation (a concept that will come up again and again). In the case of a competitive cluster, the winning node has an output value of 1, and the others are all zero. Thus the Hebbian learning rule becomes:

$$\Delta w_{kj}(n) = \begin{cases} \alpha\left[x_j(n) - w_{kj}(n)\right] & \text{if neuron k wins} \\ 0 & \text{if neuron k loses} \end{cases}$$

Only one neuron (or cluster in biology) learns at each stage and its weights move toward the location of the input. Thus, the individual nodes specialize on sets of similar patterns and become feature detectors. Competitive learning is typically used for clustering or vector quantization. Hebbian learning is used widely throughout the neural

network field, but in its simplest form is often used for principal component analysis.

Kohonen SOMs

The Kohonen map or self-organizing feature map (SOM) is a neural network

inspired by sensory mappings commonly found in the brain [Wil76][Koh82]. A self-

organizing feature map creates a topographic map of the input patterns, in which the

spatial locations of the neurons in the lattice correspond to intrinsic features of the input

patterns. In this structure, neurons are organized in a lattice where neighboring neurons

respond to similar inputs. The result of mapping similar inputs to neighboring outputs is

a global organization that is extracted from the local neighborhoods. Topographical

computational maps have been found in many locations in the brain including the vision

areas (angle of tilt of line stimulus, motion direction), auditory areas (representations of

frequency, representations of amplitude, representations of time intervals between

acoustic events) and in motor control areas (control of eye movements). More abstract

topographic maps have been found in other parts of the brain. For example there is a map

for the representation of the location of a sound source based on the interaural differences

in an acoustic signal. The SOM is one of the most widely used unsupervised artificial

neural network algorithms.[Kan94]

The typical SOM is composed of an input layer and an output layer as shown in

Figure 2-7. The input layer broadcasts the vector input to each node in the output layer,

scaled by the weights of each connection. Each node has an input term and lateral

feedback term. The topographic mapping is created by the local lateral feedback, where neighboring connections are excitatory and more distant connections are inhibitory. This is called a "mexican hat" lateral connectivity and is shown in Figure 2-6. The result is similar to the standard competitive network except that the network creates a more gentle cutoff, thus creating a Gaussian shaped output after the lateral interconnections have stabilized. This is called a "soft-max" rule (or soft-competition) where the winning PE and a few "near-winner" PEs remain active. The competitive rule is called a "hard-max" rule, hard competition, or winner-take-all rule. Depending on the characteristics of the mexican hat lateral interconnections, the resulting output will be a gaussian of varying widths centered roughly at the location of the maximum output. The process can be described using the following equations

$$y_j = \varphi\left(I_j + \sum_{k=-K}^{K} c_{jk} y_{j+k} \right)$$

$$I_j = \sum_{l=1}^{p} w_{jl} x_l$$

where y_j is the output of the j'th node, I_j is the input to the j'th node scaled by the weights into the j'th node, c_{jk} is the lateral weights which were described above as the mexican hat function, and ϕ is a nonlinear saturating function which keeps the nodes from growing without bound.

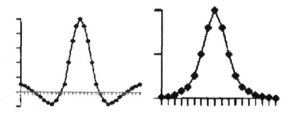

Figure 2-6: Mexican hat lateral connectivity and Gaussian shaped output

34

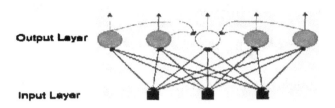

Figure 2-7: Connectivity of an SOM

After the outputs have stabilized, the network can be trained with a simple
Hebbian like rule to train the weights of the winning node and its neighbors. The
neighboring nodes can be trained in proportion to their activity (Gaussian), or all
neighbors within a certain distance can be trained equally. The learning rule can be
described as follows:

$$w_j(n+1) = w_j(n) + \eta(n)\pi_{j,i(x)}(n)[x(n) - w_j(n)]$$

where w_j are the weights of node j, $x(n)$ is the input at time n, $\pi_{j,i(x)}$ is the neighborhood
function centered around the winning node $i(x)$, and $\eta(n)$ is the learning rate. Notice that
both the learning rate and neighborhood size are time dependent and are typically
annealed (from large to small) to provide the best performance with the smallest training
time.

A simplified approximation to this algorithm consists of two stages: first, find the
winning node (the one whose weights are closest to the input), then update the weights of
the winner and its neighbors in a Hebbian manner.

The SOM is an unsupervised network with large local connectivity, but
unsupervised networks do not typically suffer from overtraining. Because the input is
mapped onto a discrete, usually lower dimension output space, the SOM is typically used

35

as a vector quantization (VQ) algorithm. The weights of the winning node are the vector quantized representation of the input.

A typical example of an SOM is mapping a two-dimensional input space onto a one-dimensional SOM. Figure 2-8 shows a random distribution of points that make up the input space in two dimensions. The points are plotted such that the coordinates of the point represent the input data. When this input data is presented to the 1-D SOM, the map trains the nodes to maintain local neighborhoods in the input space. These local neighborhoods force a global ordering of the output nodes. After training, the nodes of the SOM are ordered and the weights of the nodes represent the center of mass of the input space to which they respond. By plotting the weights of the SOM PEs onto the input space, one can see where the center of each VQ cluster is located. The SOM is more than just a clustering algorithm. It also orders the PEs such that neighboring PEs respond to neighboring inputs. To show this, we connect neighboring PEs with a line. The right side of Figure 2-8 shows how the SOM maps a one-dimensional structure to cover the two-dimensional input space. This clearly shows how the global ordering has occurred

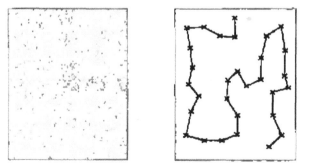

Figure 2-8: Example of a 1-D SOM mapping a 2-D input

and that the 1-D output snakes its way through the input space in order to maintain its topographic ordering and still cover the input space.

Neural Gas

The neural gas algorithm is similar to the SOM algorithm without the imposition of a predefined neighborhood structure on the output PEs. The neural gas PEs are trained with a soft-max rule, but the soft-max is applied based on the ranking of the distance to the reference vectors, not on the distance to the winning PE in the lattice. The neural gas algorithm has been shown to converge quickly to low distortion errors which are smaller than k-means, maximum entropy clustering or the SOM algorithm [Mar93]. It has no predefined neighborhood structure as in the SOM and for this reason works better on disjoint or complicated input spaces

Martinetz et. al. [Mar93] showed an interesting parallel between most of the major clustering algorithms. The main difference between each one is how the neighborhood is defined. For K-means clustering, there is no neighborhood, only the winner is trained. This is a hard max.

$$\Delta w_i = \varepsilon * \delta_{ii}(x) * (x - w_i)$$

For maximum entropy clustering, the neighborhood is defined as a soft max based on the distance in an entropy space

$$\Delta w_i = \varepsilon * h(\|x - w_i\|) * (x - w_i)$$

For the SOM, the neighborhood is based on the position in the SOM lattice

$$\Delta w_i = \varepsilon * h_\sigma(i, l(x)) * (x - w_i)$$

37

and for the neural gas algorithm the softmax is based on the ranking of the node. For

instance, the closest node gets the largest update, followed by the second closest, etc.

$$\Delta w_i = \varepsilon * h_\sigma(k_i(x,w)) * (x - w_i)$$

<u>Supervised Training</u>

Supervised training is the most commonly used and applied mechanism for

training neural networks, especially for classification and function approximation. Many

applications can be framed as a function approximation problem. The main supervised

training technique is called backpropagation [Rum86]. Typically it is applied to

multilayer perceptrons (MLPs) that consist of multiple layers of PEs, each of which does

a sum-of-products on its input and then applies a saturating nonlinearity. The

backpropagation algorithm works by first computing the forward activations of the

network by applying the inputs to the network and computing and storing the output of

every PE in the network. The activity of the network outputs is then compared to the

desired activity of the network outputs and an error is computed. This error is then

propagated backwards (thus the name backpropagation) through the network and is used

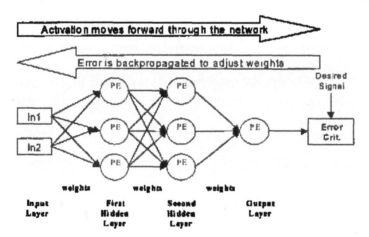

Figure 2-9: Activation and error propagation in a static neural network

to adapt each of the weights in the system. This is graphically depicted in Figure 2-9.

The output of each PE in an MLP can be described by the following equation:

$$y_j = f(net_j) = f\left(\sum_i w_{ji} x_i + b_j\right)$$

where w_{ji} represents the weight from PE i to PE j, x_i represents the output of PE i or the external input for the first layer, b_j represents the bias for PE j, and f() is the nonlinearity of the PE which is typically a logistic function (which ranges from $0 \rightarrow 1$) or a tanh function (which ranges from $-1 \rightarrow 1$). The performance surface that is searched using gradient descent is defined by:

$$J = \frac{1}{2N} \sum_{p=1}^{N} \sum_{i=1}^{m} e_{ip}^2 = \frac{1}{2N} \sum_{p=1}^{N} \sum_{i=1}^{m} (d_{ip} - y_{ip})^2$$

where e is the difference between the output and the desired signal, p is the index over the patterns and i over the output PEs. We want to update each weight based on the partial of J with respect to each weight. This formula can be written as:

$$\frac{\partial J}{\partial w_{ij}} = \frac{\partial J}{\partial y_{ip}} \frac{\partial y_{ip}}{\partial net_{ip}} \frac{\partial}{\partial w_{ij}} net_{ip} = -(d_{ip} - y_{ip}) f'(net_{ip}) x_{jp} = -\varepsilon_{ip} f'(net_{ip}) x_{jp}$$

if we define the local error δ_i for the ith PE as

$$\delta_i(n) = \frac{\partial J}{\partial y_{ip}} f'(net_{ip})$$

Then we can generalize the backpropagation algorithm for the MLP and the LMS algorithm for linear systems. All the weights in gradient descent learning are updated by multiplying the local error ($\delta_i(n)$) by the local activation ($x_i(n)$) according to Widrow's estimation of the instantaneous gradient first shown in the LMS rule

$$\Delta w_{ij}(n) = \eta \delta_i(n) x_j(n)$$

The difference between these algorithms is the calculation of the local error. If the PE is linear, then we have a linear combiner and the derivative of f is a constant. The equation then becomes the LMS rule. If the PE is nonlinear and is an output PE, then the local error is simply the difference between the output and the desired signal scaled by the derivative of the nonlinearity. This is simply the delta rule. If the PE is nonlinear and is a hidden layer PE, then the error is the sum of the backpropagated errors from the PEs that follow it.

$$\delta_i(n) = f'\big(net_i(n)\big) \sum_k \delta_k w_{ki}(n)$$

This simple rule nicely summarizes the backpropagation algorithm and shows its relationship to other adaptive algorithms.

Second Order Methods

The standard backpropagation method of training a neural network uses the LMS approximation to gradient descent, which uses only an instantaneous estimate of the gradient. Second order methods collect information over time to get a better estimate of the gradient, thus allowing for faster convergence at the cost of more computations per cycle. In linear adaptive filtering, the recursive least squares (RLS) algorithm is used for exactly this purpose. The RLS algorithm is based upon estimating the inverse of the correlation matrix of the input. With this information, the RLS algorithm can often adapt as much as ten times faster than LMS. The RLS algorithm can also be formulated as a special case of the Kalman filter. Besides faster convergence, the RLS algorithm also has two other advantages [Hay96]: the eigenvalue spread of the correlation matrix does not adversely affect the training (unlike in LMS) and the learning rate is automatically

determined (the Kalman gain).

Since RLS and Kalman filtering are derived for linear systems, they must be

modified for use with nonlinear systems. These are typically called extended RLS or

extended Kalman filtering. The most straightforward approach is to linearize the total

cost function and directly apply RLS. This requires the storage and update of the

complete error covariance matrix whose size is the square of the number of weights in the

network [Hay94]. A better approach is to apply RLS to each node individually and

linearize the activation function of the PE using a Taylor series about the current

operating point. This method is called the multiple extended Kalman algorithm (MEKA)

[Hay94] and reduces the computational requirements by ignoring the cross-terms between

PEs.

Temporal Neural Networks

As we stated previously, the majority of the temporal neural networks are

extensions of the static neural networks, either by adding memory or adding recursive

connections. This section could just as easily be called "Extending static architectures to

include time". Again, this topic will be discussed in two sections, supervised and

unsupervised neural networks.

Temporal Unsupervised Learning

This section presents the methodologies currently available to add temporal

information to unsupervised networks. Almost all work done on temporal unsupervised

training has used self-organizing maps.

As mentioned before, a self-organizing map (SOM) creates a topographic map of the input patterns, in which the spatial locations of the neurons in the lattice correspond to intrinsic features of the input patterns. In this structure, neurons are organized in a lattice where neighboring neurons respond to similar inputs. There have been many attempts at integrating temporal information into the SOM. One major technique is to add temporal information to the input of the SOM. For example, exponential averaging and tapped-delay lines were tested in [Kan90][Kan91], while coding in the complex domain was implemented in [Moz95]. Another common method is to use layered or hierarchical SOMs where a second map tries to capture the spatial dynamics of the input moving through the first map [Kan90][Kan91].

More recently, researchers have begun integrating memory inside the SOM, typically with exponentially decaying memory traces. Privitera and Morasso have created a SOM with leaky integrators and thresholds at each node which activate only after the pattern has been stable in an area of the map for a certain amount of time. This allows the map to pick out only the "stationary" regions of the input signal and use these sequences of regions to detect the input sequence [Pri93][Pri94][Pri96].

The SARDNET architecture [Jam95] adds exponential decays to each neuron for use in the detection of node firing sequences. Once a node fires for a particular sequence, it is not allowed to fire again. Therefore, at the end of the sequence presentation, the sequence of node firings can be detected (or recreated) using the decayed outputs of the SOM. The exponential decay, however, provides poor resolution at high depths and thus will perform poorly with noisy and/or long sequences.

Chappell and Taylor have created a SOM which has neurons that hold the activity on their surface via leaky integrator storage [Cha93]. This activity is added to the typical spatial distance between input and weight vector to determine the next winner. The same or neighboring nodes will thus be more likely to win the competition for successive elements in a sequence. This creates neighborhoods with sensitivity to the previous input (i.e. context). There is not a successful method available yet to train these networks. The learning law proposed by Chappel and Taylor can lead to an unstable weight space. The methodology seems to work for patterns of binary inputs with at most length 3. Critchley [Cri94] has improved the architecture by moving the leaky integration to the synapses. This gives the network a much better picture of the temporal input space and has much more stable training, but becomes nothing more than an exponentially windowed input to a standard Kohonen map, as proposed by Kangas [Kan90].

The temporal organization map (TOM) integrates a cortical column model, SOM learning and separate temporal links to create a temporal Kohonen map [Dur96]. The TOM is split into *super-units* that are trained via the SOM learning algorithm. Winning units from each super-unit fire and then decay. Temporal links are made between the currently firing node and any node which has an activity above a threshold. Thus there can be multiple links created for each firing, allowing for the pattern to skip states.

Kohonen and Kangas have proposed the hypermap architecture to include context in the SOM architecture. Kohonen's original hypermap architecture included two sets of inputs and weights [Koh91]. The first set is a context vector that is a tapped delay line of the past and future pattern vectors. This input is used to determine a "context domain" in the SOM. All nodes in the context domain are labeled active and are then presented with the current input pattern. The "pattern" weights and context weights are then trained in the typical SOM manner. Kangas extended this concept by eliminating the context weights and allowing only nodes in the vicinity of the last winner to be selected. This smoothes the trajectory of winning nodes throughout the map and allows context to affect the selection of the winner without the addition of parameters like the width of the context window [Kan92]. Kangas has also proposed an SOM architecture that has an LPC predictor at each node in the Kohonen net. This provides temporal pattern recognition by using a filter at each node where the AR filters were trained via either genetic programming or gradient descent [Kan94].

Goppert and Rosenstiel conceptually extend this concept to include the notion of attention [Gop94a][Gop94b][Gop95]. The theory being that the probability of selecting a winner is affected by either higher-cognitive processes (which may be considered a type of supervision) or by information from the past activations of the network. This gives two components to the selection of a winner, the extrasensory distance (context or higher processes) and sensory distance (normal distance form weight to input). These two components can be added or multiplied. They focus on the concept of context and create a moving area of attention, which is the region that has been activated most in the recent past. The center of attention moves as each winner is selected and the region of attention has a Gaussian weighting applied to it so that nodes near the last winner will be more likely to fire the next time. The architecture outperformed the standard SOM on simple temporal tasks but did not train well on more complicated trajectories.

Temporal Supervised Neural Networks

The main problem with temporal supervised neural networks is the complexity in training them. When the desired architecture contains recurrent connections or memory in one of the hidden layers, the network must be trained with a temporal gradient descent algorithm. There are two distinct approaches to the problem, modifying the architecture to simplify the temporal gradient calculations and creating better and/or faster methods of training the temporal neural networks.

Architectural approaches

The focused time-delay neural network (TDNN) has memory added only at the first layer and is the simplest example of an architecture designed to avoid many of the complications of temporal neural networks. It is simply a static MLP with a tap delay line between the input and the first layer. Because the memory is restricted to the first layer, the network can still be trained using static backpropagation. The tap delay line maps a segment of the input trajectory into an N-dimensional static image that is then mapped by the MLP. This works quite well for many applications, but has a number of difficulties as mentioned previously. The main difficulties are the increased number of weights required for TDNNs (each input now requires m weights where m is the number of taps in the tap delay line) and the inflexible, prewired nature of the tap delay line.

Some of the problems with TDNNs have been attacked by defining the connectivity between layers such that only certain regions of each layer are connected. By doing this, certain regions in the input layer, corresponding to certain time periods of the input, can be connected to a single region of the second layer. This provides a more goal directed architecture that can be time-shift or frequency-shift invariant. Although this can reduce the effects of the problems of TDNNs, the problems still remain and each network must be tailored for each application. [Saw91][Haf90]

Two other networks deal with temporal information by using a very restrictive type of feedback. The Jordan network [Jor86] use recurrency between the output and the input of the network. The output of the network is fed back to a context unit which is simply a leaky integrator. The Elman network [Elm90] provides feedback from the hidden layer to the context units in the input layer. This is potentially more powerful than the Jordan network because it stores and uses the past state of the network, not just the past output of the network. Although both networks are commonly found in the neural network literature, neither is particularly powerful or easily trained.

Recurrent networks are also continuously being modified in an attempt to improve their performance on temporal problems. Mozer has proposed a "multiscale integration model" that uses recurrent hidden units that have different time constants of integration, the slow integrators forming a coarse but global sequence memory and the fast integrators forming a fine grain but local memory [Moz92]. This work, however, is based only on exponentially decaying memory and the problem of selecting the time constants has not been solved (the time constants have to be hand tuned).

A different spin on recurrent networks is the use of "higher order networks". These networks are recurrent networks where:

- hidden units represent states and the output of these states are fed back and *multiplied* with the inputs of the nodes, thus allowing second order statistics to be used [Gil91][Wat91]

- one network computes the weights for a second network [Pol91][Sch92a][Sch92b]

The higher order networks have proven to be excellent sequence recognizers (grammar recognizers), but have failed to make a serious impact on temporal processing. These networks provide a representation for states in the neural network and allow the computation of high order statistics. For example, a second order network can compute the autocorrelation of the input, thus creating a translation invariant architecture. The main disadvantage of this work is that for complex tasks, higher order networks require even more weights and have even more complicated performance surfaces than standard ANNs.

Algorithmic approaches

There are two fundamental methods of computing the gradient for a dynamic neural network. First, the gradients can be computed in the backward direction similar to the static backpropagation techniques from feedforward networks. Unfolding a recurrent network in time creates a large, static, feedforward network where each "layer" consists of an instance of the recurrent network at each time step. Backpropagation can then be applied to this large feedforward network and the gradient can be computed. This is called backpropagation through time (BPTT) [Rum86]. The main shortcoming of this technique is that it is non-causal. The BPTT algorithm must be used in a batch mode, the data travels first in the forward direction while the entire state of the network is saved at each step. Next, the error is backpropagated in the reverse temporal order. A secondary shortcoming of BPTT is the memory required to store the state of the network at each iteration.

Many alterations have been made to the BPTT algorithm to improve its utility, in particular to make it usable as an on-line algorithm. Williams and Peng [Wil90] used a history cutoff where they assumed that the gradient information from the distant past is relatively inconsequential and thus can be ignored. Combining this and the use of a small step size, the new algorithm, BPTT(k) can be used in an on-line manner. See Pearlmutter [Pea95] for a review of this technique and others.

The second fundamental method of computing the gradient for a recurrent neural network computes the gradients in the forward direction. This method, called RTRL [Wil89], computes the partial of each node with respect to each weight at every iteration. The method is completely on-line and simple to implement. The main difficulty with the RTRL method is its computational complexity. If we define n to be the number of PEs and m to be the number of weights, then the computation of the gradients of each PE with respect to each weight is $O(n^2 m)$. For a fully recurrent network, this dominates the computational complexity and requires $O(n^4)$ computations per step. The algorithm works quite well on small networks, but the n^4 factor becomes overwhelming as the number of nodes increases.

The RTRL algorithm for a recurrent network can be summarized by the following set of equations [Hay94]. First, we define set A as the set of all inputs, set B as the set of all PEs, and set C as the set of outputs with desired signals. The forward activation equations are:

$$net_j(n) = \sum_i w_{ji}(n)u_i(n)$$

$$y_j(n+1) = \varphi(net_j(n))$$

where u represents the input vector at each time step and is composed of both the external inputs and the outputs of each PE (the values of the feedback). The gradient descent technique is based upon computing the sensitivity of each PE with respect to each weight. The weights are updated on-line using these sensitivities:

$$\Delta w_{kl}(n) = -\eta \frac{\partial \varepsilon(n)}{\partial w_{kl}(n)} = -\eta \sum_{j \in C} e_j(n) \frac{\partial y_j(n)}{\partial w_{kl}(n)}$$

For implementation, we create a matrix π that represents these sensitivities and write an update equation for it:

$$\pi_{kl}^j = \frac{\partial y_j(n)}{\partial w_{kl}(n)} \quad j \in B, k \in B, l \in A \cup B$$

$$\pi_{kl}^j(n+1) = \varphi'(v_j(n)) \left[\sum_{i \in B} w_{ji}(n)\pi_{kl}^i(n) + \delta_{kj}u_l(n) \right]$$

$$\pi_{kl}^j(0) = 0$$

π is a matrix of gradients with the rows representing weights and the columns representing nodes, thus it contains mn elements.

Many methods have been proposed to increase the speed of RTRL. Schmidhuber and others have mixed BPTT and RTRL which reduces the complexity to $O(nm)$ [Sch92]. This technique takes blocks of BPTT and uses RTRL to encapsulate the history before the start of each block. Sun, Chen, and Lee have developed an $O(nm)$ on-line method based on a Green's Function approach [Sun92]. By solving an auxiliary set of equations, the redundancies in the computation of the sensitivities over time can be removed. Zipser approached the problem in a different way and reduced the complexity of the RTRL algorithm by simply leaving out elements of the sensitivity matrix based upon a subgrouping of the PEs [Zip89]. The PEs are grouped arbitrarily and sensitivities between groups are ignored. If the size of the subgroups remains constant, then this reduces the complexity of the RTRL algorithm to $O(m)$. This is a tremendous improvement, however, the method lacks some of the power of the full RTRL algorithm. It sometimes requires more PEs than the standard RTRL algorithm to converge.

Second order methods

Many researchers argue that simple gradient descent [Moz94] is not sufficiently powerful to discover the sort of relationships that exist in temporal patterns, especially those that cover long time sequences or involve high order statistics. Bengio, Frasconi, and Simard [Ben93] also present theoretical arguments for the inherent limitations of learning in recurrent networks. Many researchers have recently started using the extended Kalman filter algorithm, which is very similar to the RLS algorithm, for training dynamic neural networks. As described previously, the extended Kalman filter algorithm uses information from the correlation matrix, which is accumulated over time, to better approximate the direction to the bottom of the performance surface. Again, the problem with the extended Kalman filter algorithm is that it requires the computation and storage of the correlation matrix between the weights of the system. The computational requirement for this method is $O(N^2)$. The standard method of reducing this computational load is to decouple each PE of the network, such that the correlation matrix is only computed between weights that terminate at the same PE. Puskorius and

Feldkamp [Pus94] call this method the decoupled extended Kalman filter (DEKF)

algorithm.

The main difference between the dynamic version of the RLS/EKF algorithm and

the static version is that the gradients that are used in the second order calculation are the

dynamic gradients, not the static gradients. Thus, the BPTT or RTRL algorithms must

still be used to compute these gradients.

Sequence Recognition

There are two broad categories of temporal problems which are typically

addressed in the literature: sequence recognition and temporal pattern processing.

Sequence recognition is typically a process of recognizing (and often reproducing)

discrete symbolic sequences. These problems typically focus on recognizing grammars

and symbolic patterns. Temporal pattern processing, however, involves the recognition,

identification, control or other processing of a continuous signal which varies with time.

Speech recognition is an example of temporal pattern recognition. The continuous signal

can be vector quantized and turned into a symbolic pattern, but it is not practical to then

treat it as a sequence recognition problem. Temporal patterns of interest are difficult to

accurately quantize and typically have various forms of time warping and noise which

make sequence recognition of quantized temporal patterns nearly impossible.

Since the emphasis of this proposal is not sequence recognition, we will only

briefly introduce a few interesting neural networks which accomplish this task. Wang

and Arbib have proposed models based on the two dominant theories of "forgetting" – the decay theory of forgetting where the memories decay from the time they are entered [Wan90] and the interference theory of forgetting where memory only decays when new inputs which must be remembered arrive [Wan93]. Both architectures are based on a winner-take-all field of neurons where the winning node fires and is then decremented slowly. The sequence is detected using an extra "detector unit" which is trained by the Hebbian rule using *attentional learning*. The main difficulty with these and other sequence recognizers is that they tend to be intolerant of time-warping and missing or noisy data – problems which are prevalent in temporal pattern recognition.

The outstar avalanche was an early neural network that was used to learn and generate temporal patterns [Gro82]. The outstar avalanche is composed of N sequential outstars which detect an input and each outstar triggers the next in a chain producing an avalanche effect. This architecture was modified to include the combined effect of the input dot product and the avalanche input from preceding nodes and was called the spatio-temporal network (STN) [Fre91]. The sequential competitive avalanche field (SCAF) [Hec86] is a further extension of the STN where each node has lateral interconnections allowing the outstars to be competitive.

Comparison of Hidden Markov Models with ANNs

Due to the difficulties in modeling sequential structure with ANNs, hidden Markov models have become the gold standard for modeling many temporal processes (e.g. speech). Time sequence matching is a major problem in applying neural nets to temporal/dynamical, non-stationary processes. Although ANNs have been successfully applied to time series prediction [Wei94], they have not been as successful in tasks that have synchronization problems such as time-warping. For example, different utterances of the same word can have very different timescales; both the overall duration and the details of timing can vary greatly. ANN models for speech have been shown to yield good performance only on short isolated speech units (e.g. phoneme detection). They have not been shown to be effective for large-scale recognition of continuous speech. The TDNN, for example, has powerful methods for dealing with local dynamic properties, but cannot deal with sequences explicitly.

The HMM provides a compact, tractable mechanism for handling this temporal information by including explicit state information. Various neural network techniques have attempted to add state information, typically via feedback, but have been only successful on modest size applications. HMMs are stochastic in nature and thus can succeed even when the temporal nature of the system is locally very noisy. Speech patterns, for example, are to some extents a sequential process, however, they are sufficiently ambiguous locally that it is not adequate to make decisions locally and then process sequences of symbols.

Two formal assumptions characterize HMMs as used in speech recognition. The first-order Markov hypothesis states that history has no influence on the chain's future evolution if the present is specified – e.g. the temporal information is stored in the current state of the system and all relevant temporal information must be able to be stored in this way (there is no other memory in the system). The second assumption is that the outputs depend stochastically only on the state of the system.

The two main advantages of ANNs over HMMs is that ANNs are discriminative and ANNs do not rely on the Markov assumptions. Typically HMMs are trained using a within-class method (each model is trained only on in-class, segmented, data). ANNs, however, can be trained to find the differences between classes, thus they can discriminate between classes, not just detect/model classes. ANNs have few restrictions on the systems they can model. The HMMs, however, assume that the observations are independent and that the underlying process that is modeled is a Markov Process. New methods which marry the discriminative power of the ANN with the temporal nature of the HMM have been relatively successful [Bou90].

CHAPTER 3
TEMPORAL SELF-ORGANIZATION

Introduction and Motivation

As described in the previous chapters, working with temporal patterns has been a very difficult task for neural networks. This problem is largely due to the fact that the methodologies applied to temporal processing are simple extensions of static neural networks with little regard for the unique nature of time and time based signals. Most of these architectures simply add memory to a well-known static network and can achieve reasonable performance for simple problems, but do not perform as well on more complex problems. Like in the 1980s when pattern recognition and classification drove the research community to develop neural networks, biological systems still easily outperform state-of-the-art solutions to temporal processing problems. For this reason, I began researching biological neural networks and biological mechanisms that might help us better solve these problems.

As my research progressed, two key aspects continually resonated with my underlying goal of creating better neural networks for temporal pattern processing. These two elements are the self-organization of similar or correlated cells into clusters or neighborhoods (similar to place cells in the Hippocampus), and the diffusion of information over time and space. Self-organization describes a system where each individual entity in the system has only simple local rules regarding its behavior. These simple local rules, however, can create global organization without any global control. Self-organization applies at virtually every layer of the universe, from neurons and brain cells, to bug populations, to solar systems and galaxies. It is tremendously important in the formation of the brain and in my opinion is greatly underutilized in artificial neural networks.

The second element is diffusion. Like self-organization, diffusion is found everywhere. It can be derived from simple random Brownian motion (simple local rules as well), where particles and other objects move from areas of large densities to areas of small densities. Diffusion itself is a rather simple concept that may not appear to add much to neural network theory. However, when you add diffusion to a dynamical system (for instance, the reaction-diffusion equations), the resulting system can obtain some tremendously interesting and powerful dynamics.

The Model

Most temporal neural networks use short-term memory to transform time into space. This time-to-space mapping is usually the only mechanism for dealing with temporal information. The neural network operates as if the temporal pattern was simply a much larger spatial pattern. This is clearly inefficient. My method uses *diffusion* to create *self-organization in time and space*. The theory is to leave the fundamentals of the neural network the same (in order to use the theory and knowledge we have already accumulated) but to add self-organization in *space-time* to the PEs in the network. By creating temporally correlated neighborhoods in the field of PEs making up the network, the basic functionality of the network is more organized and temporally sensitive, without drastically changing its underlying operation. The mechanism for the creation of these temporally correlated neighborhoods is the diffusion mechanism. In the brain, NO is given off from firing neurons in the brain, and diffuses throughout. NO has also been shown to affect the sensitivity of the neuron to synaptic changes (e.g. weight changes in neural networks). It has been theorized that this diffusion of NO may be responsible for the creation of place cells and other organization in the brain.

In a more abstract sense, the diffusion of NO can be considered the diffusion of the *neural activity*. When a large group of neurons fire in close proximity (both temporally and spatially), a local build-up of NO probably occurs and diffuses throughout the brain. In my architectures, I use the concept of activity diffusion to create the temporally correlated neighborhoods. When a PE or group of PEs fire, they influence their neighbors, typically lowering their threshold such that they are more likely to fire in the near future. Because the underlying mechanism of most neural network training is Hebbian in nature, when neighboring PEs fire in a correlated fashion, they tend to continue to fire in a correlated fashion. This creates the temporally correlated neighborhoods and the self-organization in space-time.

I have applied this concept to three different ANN architectures. The first is based on the self-organizing map (SOM) and is the most biologically inspired. The second is based on the neural gas algorithm which provides a more powerful but functionally similar solution. Lastly, to prove the robustness of the method, I applied it to the training of recurrent MLPs. MLPs are a totally different architecture and are trained in a totally different manner (e.g. supervised vs. unsupervised training). The MLP is not biologically relevant, but the temporal self-organization method still proved to decrease training times dramatically. The rest of this chapter is divided into three sections based on these architectures. It is arranged chronologically so that the presentation will be more smooth, even though the MLP architecture may be the most useful of the three. This chapter only presents the theoretical derivation of each architecture and a simple illustrative example for each. Detailed application of each method to more practical problems will be presented in the next chapter.

Temporal Self-Organization in Unsupervised Networks

This section describes the two unsupervised networks which I have applied the concept of temporal clustering. The first architecture is based on the self-organizing map and is called the self-organizing temporal pattern recognizer (SOTPAR). The second architecture is based on the neural gas algorithm and is called the SOTPAR2.

Temporal Activity Diffusion Through a SOM (SOTPAR)

The self-organizing temporal pattern recognizer (SOTPAR) [Eul96a][Eul96b] is a biologically inspired architecture for embedded temporal pattern recognition (finding patterns in an unbounded input sequence without segmentation or markings). This is a difficult task since the patterns must be searched from every possible starting point. Although the SOTPAR architecture is unsupervised and thus cannot be used efficiently as a pattern recognition device, it does preprocess the input such that patterns commonly found in the training data will be easily detectable from the output of the SOTPAR. Most of the emphasis in this work is in the proper temporal representation of the spatio-temporal data.

The SOTPAR architecture adds two temporal characteristics to the SOM architecture, activity diffusion through the space of output PEs and the temporal decay of activations. Using these concepts, the SOTPAR converts and distributes the temporal information embedded in the input data into spatial connections and ordered PE firings in the network, all using self-organizing principles.

Similar to self-organizing maps, the network uses competitive learning with neighborhood functions [Koh82]. In the SOM, the input is simultaneously compared to the weights of each PE in the system and the PE that has the closest match between the input and its stored weights is the winner. The winner and its neighbors are then trained in a Hebbian manner, which brings their weights closer to the current input.

The key concept in the SOTPAR architecture is the activity diffusion through the output space. The firing of a PE in the network causes activity to diffuse through the network and affects both the training of the network and the recognition of the network. In the SOTPAR, the activity diffusion moves through the lattice of an SOM structure and is modeled after the reaction-diffusion equation [Mur89]

$$\frac{\partial m_i(x,t)}{\partial t} = f(m_i(x,t), m_j(x,t)) + D_{m_i} \frac{\partial^2 m_i(x,t)}{\partial x^2}$$

where m_i can be considered the activity of PE i, $f(*)$ can be considered the current match, and the second derivative is the diffusion of activity over space and time. If the system is "excitable media" (multi-stable dynamical system), then the diffusion of activity can create traveling pulses or wavefronts in the system. When the activity diffusion spreads to neighboring PEs, the thresholds of these neighboring PEs are lowered, creating a

59

situation where the neighboring PEs are more likely to fire next. I define enhancement as the amount by which a PE's threshold is lowered.

In the SOTPAR model, the local enhancement acts like a traveling wave. This significantly reduces computation of diffusion equations and provides a mechanism where temporally ordered inputs will trigger spatially ordered outputs. This is the key aspect of this network architecture. The traveling wave decays over time because of competition for limited resources with other traveling waves. It can only remain strong if spatially neighboring PEs are triggered from temporally ordered inputs, in which case the traveling waves are reinforced. In a simple one dimensional case, Figure 3-1 shows the enhancement for a sequence of spatially ordered winners (winners in order were PE1,

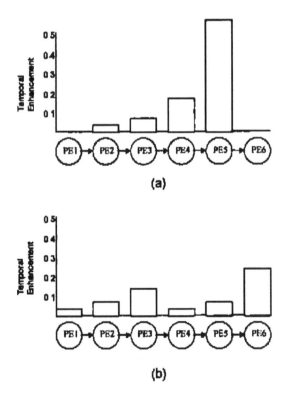

Figure 3-1: Temporal activity in the SOTPAR network. a) activity created by temporally ordered input; b) activity created by unordered input

PE2, PE3, PE4) and for a sequence of random winners (winners in order were PE4, PE2, PE1, PE5), which would be the case if the input was noise or unknown. In the ordered case, the enhancement will lower the threshold for PE 5 dramatically more than the other PEs making PE 5 likely to win the next competition. In the unordered case, the enhancement becomes weak and affects all PEs roughly evenly.

The second temporal functionality added to the SOM is the decay of output activation over time. This is also biologically realistic [Cha93]. When a PE fires or becomes active, it maintains an exponentially decaying portion of its activity after it fires. Because the PE gradually decays, the wavefront it creates is more spread out over time, rather than a simple traveling impulse. This spreading creates a more robust architecture that can gracefully handle both time-warping and missing or noisy data. The decay of the activity also creates another biological possibility for explaining the movement of the enhancement throughout the network. If we define a neighborhood around a neuron as one where it has strong excitatory connections with its neighbors, then the decay of activity from a neuron which fired in the past will help to fire (or lower the threshold of) its neighboring PEs.

Algorithm description

To simplify the description of the algorithm, I will use 1D maps and let the activity propagate in only one direction, since the diffusion of the activity is severely restricted in the one-dimensional case. Thus, the output space can be considered a set of PEs connected by a string where the information is passed between PEs along this string. The activity/enhancement moves in the direction of increasing PE number and decays at each step. An implementation of the activity diffusion in one string is shown in Figure 3-2 and includes the activity decay at each PE and the activity movement through the net in the left-to-right direction. The factors u and $(1-u)$ are used to normalize the total activity in the network.

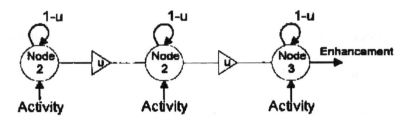

Figure 3-2: Model for activity diffusion in one string of the SOTPAR

This activity diffusion mechanism serves to store the temporal information in the network. During training, the PEs will be spatially ordered to sequentially follow any temporal sequences presented. At each iteration, the activity of the network is determined by calculating the distance between the input and the weights of each PE and allowing for membrane potential decay:

$$act(t,x) = act(t-1,x)*(1-u) + dist(inp(t), w_x)*(u)$$

where $act(t,x)$ represents the activity at PE x at time t, and $dist(inp(t), w_x)$ represents the distance between the input at time t and the weights of PE x. Typically the activity is thresholded and enhanced before being propagated. For example

$$act' = \max(act-.5,0)*2$$

next, the winning PE is selected by

$$winner = \arg\max(act + \beta * enhancement)$$

where the enhancement is the activity being propagated from the left. The parameter β is the *spatio-temporal parameter* that determines the amount that a temporal wavefront can lower the threshold for PE firing. By increasing β you can lower the threshold of neighboring PEs to the point where the next winner is almost guaranteed to be a neighbor of the current winner and forces the input patterns to be sequential in the output map. It is interesting to note that as $\beta \rightarrow 0$, the system operates like a standard SOM and when $\beta \rightarrow \infty$ the system operates like an avalanche network. [Gro82]

Once the winner is selected, it is trained along with its neighbors in a Hebbian manner with normalization as follows:

$$w_x = w_x + \eta * neigh(x) * (inp(t) - w_x)$$

where the neighborhood function, *neigh(x)*, defines the closeness to the winner (typically a Gaussian function), and the learning rate is defined by η. In our current implementation, the spatio-temporal parameter, the learning rate, and neighborhood size are all annealed for better convergence.

Representation of memory

The activity diffusion in this network creates a unique spatio-temporal memory that stores and distributes the temporal information in the network itself. Most short-term memory structures can be described by convolving the input sequence with a kernel that describes the structure of the memory. This kernel is typically one-dimensional and describes the temporal features of the memory, i.e. the depth of the memory. The SOTPAR's memory is implemented in its "enhancement" which moves through time and space. Thus, the SOTPAR memory kernel is spatio-temporal and must be described in at least two dimensions.

There are two slightly different ways to implement the temporal enhancement in a 1D SOTPAR. The difference lies in the decaying exponential portion. In method number 1, only the activity at each node is decayed. The contributions from the wavefronts do not contribute to the time dependent behavior of each node. The equation for this system is:

$$E(n,t) = E(n-1, t-1)^* \mu + A(n,t)$$
$$A(n,t) = A(n,t-1)^* (1-\mu) + In(n,t)$$

where $E(n,t)$ is the enhancement at node n at time t, $A(n,t)$ is the activity at node n at time t, and $In(n,t)$ is the matching results between the input and the weights of node n and time t. Expanding these equations gives the following results:

$$\begin{aligned}
E(n,t) &= E(n-1, t-1)^* \mu + A(n, t-1)^*(1-\mu) + In(n,t) \\
&= \big(E(n-2, t-2)^* \mu + A(n-1, t-2)\big)^* \mu + A(n, t-2)^*(1-\mu)^2 + In(n, t-1)^*(1-\mu) + In(n, t \\
&\vdots \\
&= \sum_{k=0}^{n}\sum_{\tau=0}^{n} In(n-k, t-k-\tau)\mu^k (1-\mu)^\tau
\end{aligned}$$

This equation shows how the results from the matching activity (which is called "input", for lack of a better word) contribute to the enhancement. The traveling waves create two decaying exponentials, one which moves through space (μ^k), and one which moves through time ($(1-\mu)^\tau$). The past history of the node is added to the enhancement via the recursive self-loop in $(1-\mu)$. The wavefront motion is added to the enhancement via the diagonal movement through the left-to-right channel scaled by μ. The farther the node is off the diagonal and the farther back in time, the less influence it has on the enhancement.

The SOTPAR enhancement equation is similar to the gamma memory impulse response for tap n:

$$g_n(t) = \binom{t-1}{n-1} \mu^n (1-\mu)^{t-n} U(t-n)$$

By doing a variable substitution, the τ can be replaced with $t\text{-}n$ in the SOTPAR equation making the two equations even more similar. The SOTPAR enhancement, however, is not an impulse response equation. The SOTPAR allows input at each element of the memory structure, unlike the gamma memory which is a generalized tapped delay line, thus the input at different times and spatial locations is required to describe the enhancement (i.e. an impulse response does not represent the desired information). In summary, the SOTPAR enhancement is a spatially distributed gamma memory with inputs at each tap.

The second method for implementing the enhancement is to allow the enhancement to also pass through the self-feedback at each node. This will allow an input to add to the enhancement multiple times by following different paths in the network. For example, *In(n-1,t-2)* can reach *E2(n,t)* either by looping first at node *n-1* and then moving to position *(n,t)* or by first moving to position *(n,t-1)* and then looping at node *n* until *(n,t)*. The equation for the enhancement in this case is:

$$E2(n,t) = E2(n-1,t-1)*\mu + E2(n,t-1)*(1-\mu) + In(n,t)$$

$$= \big(E2(n-2,t-2)*\mu + E2(n-1,t-2)*(1-\mu) + In(n-1,t-1)\big)*\mu +$$

$$\big(E2(n-1,t-2)*\mu + E2(n,t-2)*(1-\mu) + In(n-1,t-1)\big)*(1-\mu) + In(n,t)$$

$$\vdots$$

$$= \sum_{k=0}^{n}\sum_{\tau=0}^{n} In(n-k,t-k-\tau)\mu^{k}(1-\mu)^{\tau}(\tau+1)$$

This method of enhancement increases the contribution of the off-diagonal elements via the term $(\tau+1)$ and allows more flexibility in non-sequential node firings.

The two enhancement techniques can be shown for two values of μ in Figure 3-3 and Figure 3-4. Both figures show Enhancement method 1, Enhancement method 2, and the difference between Enhancement method 2 and method 1, which shows the increased influence of the off-diagonal elements.

These two figures also illustrate the effect of μ on the enhancement. With $\mu = 1$, the time decay at each node is disconnected and the enhancement moves only from node to node. With $\mu = 0$, the spatial movement of the enhancement is disconnected and only

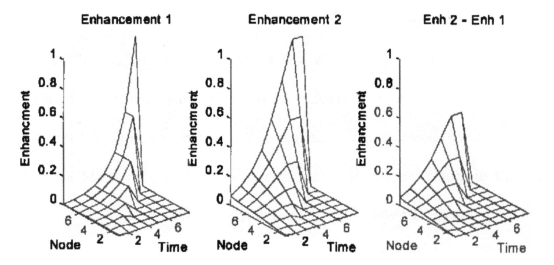

Figure 3-3 - Enhancement in the network with $\mu = 0.5$

node decay contributes to the enhancement. Lower values of μ create a broader

enhancement while higher values of μ create narrower enhancement waves where almost

all of the activity moves from one node to the next (down the diagonal of time and space).

This can be seen in the figures as a much sharper contribution to the enhancement along

the diagonal as μ moves from .5 to .75.

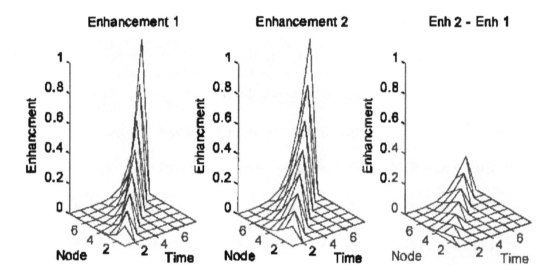

Figure 3-4 - Enhancement in the network with μ = 0.75

Another possible approach is to decouple the two exponentials μ and (1-μ). This

would require external normalization to keep the enhancement from growing without

bound, but will provide more flexibility.

A simple illustrative example

A simple, descriptive test case involves an input that is composed of two-

dimensional vectors randomly distributed between 0 and 1. Embedded in the input are 20

'L' shaped sequences located in the upper right hand corner of the input space (from

[0.5,1.0]→[0.5,0.5]→ [1.0,0.5]). Uniform noise between -0.05 and 0.05 was added to the target sequences. When a standard 1D SOM maps this input space, it maps the PEs without regard to temporal order, it simply needs to cover the 2D input space with its 1D structure. To show how this happens, we plot an 'X' at the position in the input space represented by the weights of each PE (remember, the weights of each PE are the center point of the Voronoi region that contains the inputs that trigger that PE). Since the neighborhood relationship between PEs is important, we connect neighboring PEs with a line. In a 1D SOM, the result is a "string" of PEs, and this string of PEs is stretched and manipulated by the training algorithm so that the entire input space is mapped with the minimum distortion error while maintaining the neighborhood relationships (e.g. the string cannot be broken). The orientation of the output is not important, as long as it covers the input with minimal residual energy. A typical example is shown on the left side of Figure 3-5. Note the slightly higher density of the input in the 'L' shaped region.

When the SOTPAR temporal activity is added to the SOM, the mapping has the additional constraint that temporal neighbors (sequential winners) should fire

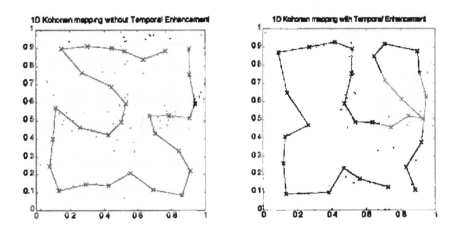

Figure 3-5 - One-dimensional mapping of a two-dimensional input space, both with and without spatio-temporal coupling

68

sequentially. Thus, the string should not only cover the input space, but also follow

prevalent temporal patterns found in the input. This is shown on the right side of Figure

3-5. Notice in the figure that sequential nodes have aligned themselves to cover the L

shaped temporal patterns found in the input.

Although not the main goal in creating the Spatio-Temporal SOM, recall is

possible after the first few samples of the sequence have been input to the network. The

rest of the pattern can be determined by following the sequence of nodes in the SOM,

although the length of the sequence is not readily determined by the map.

With a single string, the network can be trained to represent a single pattern or

multiple patterns. Multiple patterns, however, require the string to be long. A long string

may be difficult to train properly since it must weave its way through the input space,

moving from the end of one pattern to the beginning of the next. Additional flexibility

can be added by breaking up the large string into several smaller strings. Multiple strings

can be considered a 2D array of output nodes with a 1D neighborhood function. This

allows the network to either follow multiple trajectories or long complicated trajectories

in a simplified manner.

Figure 3-6 shows an example of the storage of two temporal patterns with two

strings. The left plot shows the input space that consists of two-dimensional input

vectors. Two 8-point temporal patterns (diagonal lines: bottom-left to top-right and

bottom-right to top-left) are intermixed with random noise in the input. The diagonal

lines are drawn in for clarity. Between each pattern, there is random noise. This problem

can be thought of as a motion detection problem across a visual topographic map. A

number of strings could be trained to detect motion in a variety of directions and

69

orientations. On the left side of Figure 3-6, the trained strings are shown as sequences of

8 PEs represented as 'X's (the 'O' PE denotes the beginning of the string), with

neighboring PEs connected by lines. As one can see from this figure, the memory

structure was able to extract the predominant temporal features of the input data.

The right side of Figure 3-6 shows a graphical representation of the sequence of

winning PEs after training. The horizontal axis is time, and the vertical axis is the number

of the winning PE. The input signal is labeled along the top of the plot. This plot clearly

shows that the patterns elicit the network to respond with sequential PE firings (smooth

diagonal lines), whereas the random noise between patterns causes random output firings.

Notice also that the temporal information is crucially important in the training of the

memory, especially at the center of the figure where the next point could be in one of two

possible directions. This ambiguity is responsible for the misalignment of the PEs near

the center of the input space.

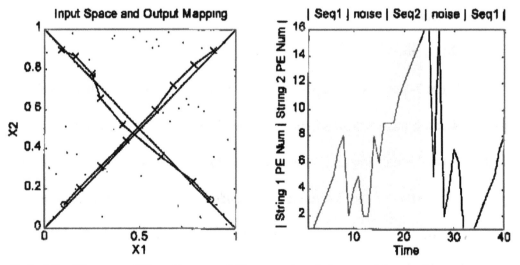

Figure 3-6 - The Storage of Two Temporal Patterns in a Memory Network

Figure 3-7 shows an example of how the network gracefully handles time

warping. In this example, the input was as in the previous example except that the target

sequences were warped to length 6, 8, and 10. The network mapped two 6 PE strings to

the diagonal targets as shown in the left side of the figure. The right side of the figure

shows the winning nodes with the three different size sequences – the first two are length

6, the second two are length 8, and the last one is length 10. The strings stretch to cover

the entire pattern and certain PEs fire more than once for a longer sequence, thus

extending the time that can be covered by the string. In general, if the network is trained

with time-warped data, it will tend to represent the target trajectories with the minimum

number of nodes (shortest pattern). The network will still respond to longer patterns by

having certain nodes win multiple times.

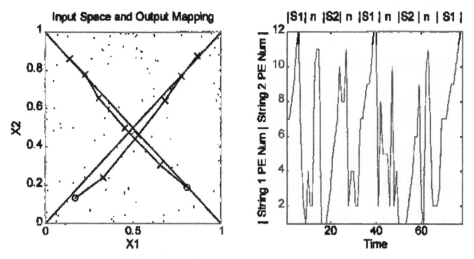

Figure 3-7 - Time Warping: Diagonal Targets Covered by Smaller Sequences

The left and middle plots of Figure 3-8 show the traveling activity over time and

space for the above example. The left side shows the activity for string 1 and the center

shows the activity for string 2. These two plots clearly show how the traveling activity builds up and reinforces the sequential firing of the output PEs (i.e. when a target

sequence is presented, the activity builds up and moves along the string). The right-side

of Figure 3-8 shows the maximum traveling activity for string 1 (solid) and string 2

(dashed). In a simple system, this plot shows how a simple threshold on the traveling

activity could be used to detect the target sequence.

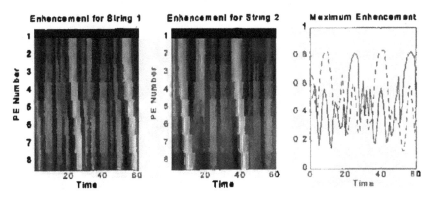

Figure 3-8: Plots of ehancement over time for string 1 and string 2 and also the maximum enhancement over time for both strings

SOTPAR summary

The SOTPAR methodology creates an array of PEs that self-organizes in space-

time with the help of temporal information. The system is trained in an unsupervised

manner and self-organizes so that sequences seen during training are mapped into unique

spatial sequential firings of the PEs at the output. The output space is similar to a

topographic map except that it maps both the temporal and spatial information. The

network embeds the temporal and input data into one output space with both temporal

and spatial locality. Instead of the standard time-to-space mapping produced by most

short-term memories, the SOTPAR produces a time-to-"time and space" mapping. The

representation is distributed throughout the self-organizing network and is stored not only in the activations of the PEs but also in the connectivity and weights of the PEs. It is a radical departure from typical neural network architectures with memory, but is actually more biologically plausible.

The SOTPAR is a unique combination of short-term and long-term memory. It contains short-term memory because the activations of the network can be used to represent a general input sequence. The interesting part of the SOTPAR, however, is that it contains attributes of a long-term memory. It stores commonly found input patterns into the network weights and produces enhanced responses to these temporal inputs. The known sequences produce an ordered response in a specific area of the output space. This is a discriminant mapping because only known sequences produce an ordered response. The sequential firing facilitates the recognition of temporal patterns by subsequent processing layers. It can also gracefully handle time warping.

Temporal Activity Diffusion In the Neural Gas Algorithm (SOTPAR2)

The SOTPAR2 network was developed to overcome a few difficulties with the original SOTPAR network. The main difficulty with the SOTPAR is the SOM map that it is built upon. The SOM's neighborhood lattice structure restricts both the movement of a trajectory through the output space of the network (e.g. the distance between successive inputs) and also limits the number of neighbors for each PE. For these reasons the neural gas algorithm is used as the basis for the SOTPAR2 architecture.

The neural gas algorithm is similar to the SOM algorithm without the imposition of a predefined neighborhood structure on the output PEs. The neural gas PEs are trained with a soft-max rule, but the soft-max is applied based on the ranking of the distance to the reference vectors, not on the distance to the winning PE in the lattice. Since the Neural gas algorithm has no predefined structure, each PE acts relatively independently. This is how it derived its name, each PE is like a molecule of gas that all spread to evenly cover the desired space. Since it has no predefined structure for the activity diffusion to move through, it allows the flexibility to create a diffusion structure that can be trained to best fit the input data. The SOTPAR2 diffuses activity through a secondary connection matrix that is trained with temporal Hebbian learning. This flexible structure decouples much of the spatial component from the temporal component in the network. In the SOTPAR, two neighboring nodes in time also needed to be relatively close in space in order for the system to train properly (since time and space were coupled). This is no longer a restriction in the SOTPAR2. This is still a space-time mapping, but now the coupling between space and time is directly controllable.

The most interesting concept that falls out of this structure is the ability for the network to focus on temporal correlations. Temporal correlation can be thought of as the simple concept of anticipation. The human brain uses information from the past to enhance the recognition of "expected" patterns. For instance, during a conversation a speaker uses the context from the past to determine what they expect to hear in the future. This methodology can greatly improve the recognition of noisy input signals such as slurred or mispronounced speech.

SOTPAR2 - algorithm details

Based on previous experience (training), the SOTPAR2 algorithm uses temporal information to lower the threshold of PEs that are likely to fire next. The standard neural gas network is appended with a connection matrix that is trained using temporal Hebbian learning. These secondary weights are similar to transition probabilities in Hidden Markov Models (HMM) and are the pathways used to diffuse the temporal information. As in the SOTPAR, the temporal activity diffusion is used to alter the selection of the winning PE and affects both the training and the operation of the network.

The SOTPAR2 algorithm works as follows: First, you calculate the distance (d_i) from the input to all the PEs. The temporal activity in the network is similar to the SOTPAR diffusive wavefronts except that the wavefronts are scaled by the connection strengths between PEs. Thus, the temporal activity diffuses through the space defined by the connection matrix as follows:

$$a_i(t+1) = \alpha a_i(t) + \frac{\sum_k \left[(\mu f(d,k) + (1-\mu)a_k(t))p_{ki} \right]}{\max(p)}$$

where $a_i(t)$ is the activity at PE i at time t, α is a decay constant less than 1, p_{ij} is the connection strength from PE i to PE j, d is the vector of distances from the input to each PE, μ is the parameter which smoothes the activity giving more or less importance to the past activity in the network, and $max(p)$ normalizes the connection strengths. The function $f(d,k)$ determines how the current match (distances) of the network contributes to the activity. At the present time, my implementation implements the case where $f(d,k)$ is simply a δ function (and the summation is removed) such that only the activity from the past winner is propagated. This is similar to the Markov model where all temporal information is stored in the state itself. Unlike the Markov model, however, the previous winners affect the output activity of the current winner. Therefore, a previous winner that has followed a "known" path through the network will have higher activity and thus will have more influence on the next selection.

In the general case for the activity equation the temporal activity at each PE is affected by contributions from all other PEs. In this case the function $f(d,k)$ is typically an enhanced/sharpened version of the output and the summation is over all PEs. This allows all the activity in the network to influence the current selection. It makes the network more robust since the wavefronts will continue to propagate (but will decay rapidly) even if the selected winner temporarily transitions to an unlikely path.

The next step of the SOTPAR2 algorithm is to modify the output (competition selection criteria) of each PE by the temporal activity in the network via the following equation:

$$out_i = d_i - \beta a_i$$

where β is the *spatio-temporal parameter* that determines how much the temporal information affects the selection of the winner. This parameter should be set based upon the expected magnitude of the noise present in the system. For example, if the data is normalized $[0,1]$, then a setting of $\beta = 0.1$ allows the network to select a new winner that is at most a distance of 0.1 farther away than the PE closest to the spatial input.

To adjust the weights, we use the standard neural gas algorithm that is simply competitive learning with a neighborhood function based on an ordering of the temporally modified distance to the input.

$$\Delta w_i = \eta h_\lambda(k_i(out))(in - w_i)$$

where η is the learning rate (step size), $h_\lambda(\cdot)$ is an exponential neighborhood with the parameter λ defining the width of the exponential. $k_j(out)$ is the ranking of PE_i based on its modified distance from the input.

The connection strengths are trained using temporal Hebbian learning with normalization. Temporal Hebbian learning is Hebbian learning applied over time, such that PEs that fire sequentially enhance their connection strength. The rationalization for this rule is that PEs will remain active for a period of time after they fire, thus both current and previous winners will be active at the same time. In the current implementation, the connection strengths are updated similar to the conscience algorithm for competitive learning:

$$\Delta p_{\arg\min(out(t-1)),\arg\min(out(t))} = b$$
$$p_{i,j} = p_{i,j}\left(\frac{N}{N+b}\right)$$

The strength of the connection between the last winner and the present winner is increased by a small constant b and all connections are decreased by a fraction that maintains constant energy across the set of connections. Another possibility for normalization would be to normalize all connections leaving each PE. This method gives poorer performance if a PE is shared between two portions of a trajectory since the connection strength would have to be shared between the two outbound PEs. It does, however, give an interpretation of the connection strengths as probabilities and points out the similarity between the SOTPAR2 and the HMM.

The parameters η and λ are annealed exponentially as in the neural gas algorithm, while β takes the form of an offset sine wave. This allows the initial phases of learning to proceed without interference so that the PEs start out with an even distribution across the input space. Then the temporal enhancement reaches a peak and slowly declines for fine tuning at the end of learning.

Operation of the SOTPAR2 network

I will use an artificial example to illustrate the features of the SOTPAR2. The input for this example is 15 pairs of noisy 8-point diagonal lines from $(0,0) \rightarrow (1,1)$ and from $(1,0) \rightarrow (0,1)$. The diagonal lines have uniform noise (± 0.15 in both dimensions) added to each point (notice the distance between each noise-free point of the diagonal lines is only 0.14). There is uniform noise [0,1] interspersed between the diagonal lines (16 points between each line such that there is twice as much noise as signal). The network extracts the temporal information from the diagonal lines *without supervision, segmentation, or labeling.* A 30-PE network was trained with and without temporal enhancement (200 iterations through the data set) and the resulting PE locations are shown in Figure 3-9 and Figure 3-10 with the diagonal lines superimposed on the figures.

Notice that the temporal enhancement during training has slightly modified the positions of the PEs. The network trained with temporal enhancement has its PEs placed more consistently near the centers of the points along the diagonal lines (in particular, look at the line segment in the lower-right). The temporal training provides a portion of the improvement made by the SOTPAR2 algorithm, but the static comparison of the network is not dramatically different.

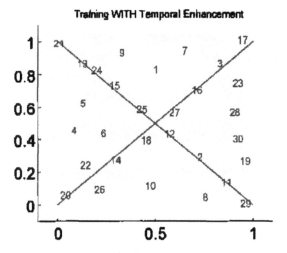

Figure 3-9: Reference vector locations after training with enhancement

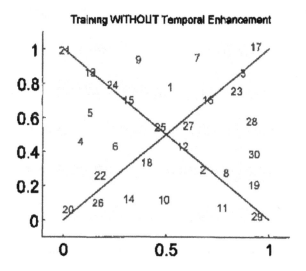

Figure 3-10: Reference vector locations after training without enhancement

During operation, *the trained weights and information from the past create temporal wavefronts in the network that allow plasticity during recognition.* This temporal activity is mixed with the standard spatial activity (distance from input to the weights) via β, the spatio-temporal parameter. Two identical inputs may fire different PEs depending on the temporal past of the signal. Figure 3-11 shows the Voronoi diagrams for the SOTPAR2 network with two different temporal histories. Voronoi diagrams

graphically describe the region in the input space that fires each PE. In these particular

diagrams, the number in each Voronoi region represents the PE number for that particular

region and is located at the center of the static Voronoi region. Remember that the center

is the same as the weights of the PE. These diagrams show the regions of the input space

that will fire each PE in the network. The left side of Figure 3-11 shows the Voronoi

diagram during a presentation of random noise to the network. Since this input pattern

was unlikely to be seen in the training input, temporal wavefronts were not created and

the Voronoi diagram is very similar to the static Voronoi diagram. The right side of

Figure 3-11 shows the Voronoi diagram during the presentation of the bottom-left to top-

right diagonal line. The temporal wavefront grew to an amplitude of 0.5 by the time PE

18 fired. Also, from the training of the network, the connection strength between PE 18

and PE 27 was large compared to the other PEs. Thus, the temporal wavefront flowed

preferentially to PE 27 enhancing its possibilities of winning the next competition.

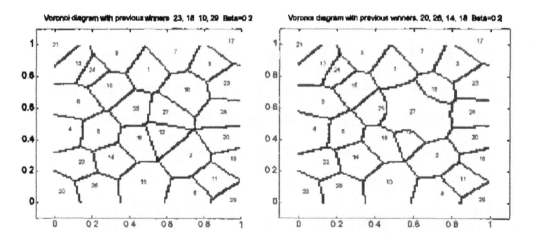

Figure 3-11: Voronoi diagrams without and with enhancement

Notice how large region 27 is in right side of Figure 3-11 since it is the next

expected winner. This plasticity seems similar to the way humans recognize temporal

patterns (e.g. speech). Notice that the network uses temporal information and its previous

training to "anticipate" the next input. The anticipated result is much more likely to be

detected since the network is expecting to see it. It is important to point out how the static

and dynamic conditions are dramatically different. In the dynamic SOTPAR2 the

centroids (reference vectors) are not as important – the temporal information changes the

entire characteristics of vector quantization creating data dependent Voronoi regions. An

animation can demonstrate the operation of the SOTPAR2 Voronoi regions much better

than static figures.

Next I created a new set of 14 noisy diagonal lines to be run through the network

as a test set. Each noisy line was passed through both a standard neural gas vector

quantization network and a SOTPAR2 VQ network. The results will be analyzed using

the 5th point in the bottom-left to top-right diagonal line. Figure 3-12 shows the locations

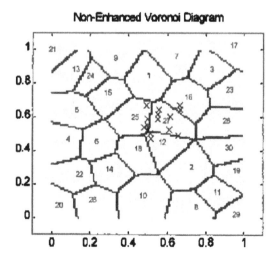

Figure 3-12: Voronoi diagram without enhancement. VQ outputs were
[12,12,16,16,25,25,25,25,27,27,27,27,27,27]

of this point in each of the 14 noisy diagonal lines along with the neural gas Voronoi

diagram. Notice that the static vector quantization cannot consistently quantize this 5th

point to the same Voronoi region. In fact, this point falls into four different regions. The

SOTPAR2 network, however, was able to quantize every one of the 5th points into the

same region. Figure 3-13 shows why. Figure 3-13 shows a typical Voronoi diagram for

the trained SOTPAR2 network after the input of the first four points of a single noisy

diagonal line. The location of the 5th point in each of the 14 noisy diagonal lines was

again plotted. Notice that now all 14 points fall into the correct Voronoi region.

Remember that each particular input sequence will create a different Voronoi diagram,

but Figure 3-13 illustrates the mechanism for the SOTPAR2's improved vector

quantization. The temporal plasticity has increased the size of the anticipated next region

and reduced the variability of the SOTPAR2 vector quantization.

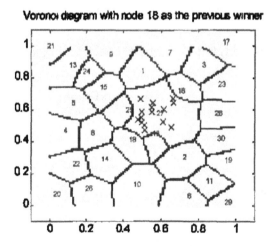

Figure 3-13: Voronoi diagram and VQ with enhancement. VQ outputs were
[27,27,27,27,27,27,27,27,27,27,27,27,27,27]

Next I ran the new noisy diagonal lines through the network and histogrammed

the VQ outputs for each point of the two lines. Figure 3-14 shows the results with the

82

point number along the horizontal axis and the node number along the vertical axis. The

number of firings for each node is indicated by the shading – white is high and gray is

low. The left-to-right diagonal line is shown in the first 8 points of the horizontal axis

and the right-to-left diagonal line is shown as the second 8 points of the horizontal axis.

Notice how much cleaner the temporal enhanced VQ output is than the standard neural

gas VQ.

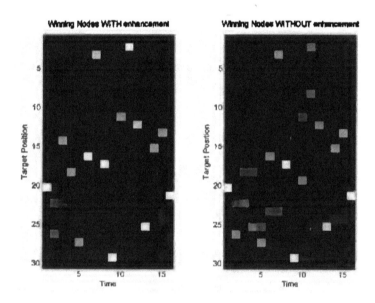

Figure 3-14: Histograms of the number of firings for each PE (bright = more) for the
networks with and without enhancement

Figure 3-15 shows a specific example of the VQ output of the two networks and

illustrates how the SOTPAR2 uses temporal information to remove noise from the input.

The input is a noisy diagonal line from bottom-right to top-left (solid line). The

SOTPAR2 output is the short dashed line, and the static VQ output is the long dashed

line. Notice how much closer the temporal VQ output is to the diagonal than the noisy

input or the output of the static VQ.

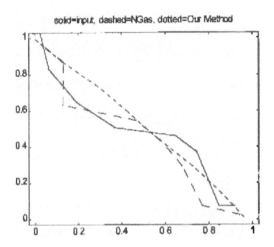

Figure 3-15: The SOTPAR2 VQ (dotted) is closer to the noise free signal than the original (solid) or the neural gas VQ (dashed)

SOTPAR2 summary

The SOTPAR2 algorithm uses temporal plasticity induced by the diffusion of activity through time and space. The SOTPAR2 algorithm is a temporal version of the neural gas algorithm that uses activity diffusion to couple space and time into a single set of dynamics that can help disambiguate the static spatial information with temporal information. This creates time-varying Voronoi diagrams based on the past of the input signal. This dynamic vector quantization helps reduce the variability inherent in the input by anticipating (based on training) future inputs.

Temporal Self-Organization for Training Supervised Networks

This section shows how the concepts of temporally trained clustering can help speed up the training of supervised neural networks. In particular, we have applied it to recurrent neural network training. Recurrent neural networks are more powerful than feedforward neural networks, but their training is very difficult and time-consuming.

Supervised neural networks are typically trained with gradient descent learning, which provides a more mathematically sound foundation than in the unsupervised networks. This allows for a goal-driven approach with mathematical derivations of the concepts. The goal of this architecture is to temporally organize the training of a recurrent neural network. A mathematical analysis will derive a principle very similar to that used in the neural gas network, that temporal correlation can be used to train PEs to form temporal neighborhoods.

Using Temporal Neighborhoods in RTRL

In the past, static neural networks and feedforward networks with memory (TDNN, etc.) have been the workhorses of the neural network world. Recently recurrent neural networks have been getting more attention, especially when applied to dynamical modeling and system identification and control. The main difficulty in training recurrent neural networks is that the gradient is a function of time. The gradient at the current time depends not only on the current input, output, and desired signal, but also on all the values in the past.

As discussed in Chapter 2, there are two fundamental methods of computing the gradient for a recurrent neural network. First, the gradients can be computed in the backward direction similar to the static backpropagation techniques from feedforward networks. This is called backpropagation through time (BPTT) [Rum86]. The main shortcoming of this technique is that it is non-causal. The second fundamental method of computing the gradient for a recurrent neural network computes the gradients in the forward direction. This method, called RTRL [Wil89], computes the partial of each node with respect to each weight at every iteration. The method is completely on-line and simple to implement. The main difficulty with the RTRL method is its computational complexity. If n is the number of PEs in a fully recurrent network, then the computation of the gradients of each PE with respect to each weight is $O(n^4)$. This algorithm can only be used for small networks.

Many methods have been proposed to increase the speed of RTRL. Zipser's approach [Zip89][Zip90] will be used here because it lends itself to our techniques. Zipser approached the problem of reducing the complexity of the RTRL algorithm by simply leaving out elements of the sensitivity matrix based upon a subgrouping of the PEs. The PEs are grouped arbitrarily and sensitivities between groups are ignored. If the size of the subgroups remains constant, then this reduces the complexity of the RTRL algorithm to $O(n^2)$. This is a tremendous improvement, however, the method lacks some of the power of the full RTRL algorithm. For example, it will sometimes require more PEs than the standard RTRL algorithm to converge.

Our methodology extends Zipser's technique by allowing the subgroups to change dynamically during learning. The dynamic subgroups are created by using an unsupervised temporal clustering very similar to that used in the SOTPAR and SOTPAR2. A derivation of a first-order approximation to the full sensitivity matrix shows that temporal correlation (temporal Hebbian learning) can be used to determine which nodes should be in each group. This method has the same computational complexity as Zipser's, but trains better and more consistently.

Review of RTRL and Zipser's Technique

The computational complexity of the RTRL algorithm is dominated by the need to update a large array of sensitivities at each step of the algorithm. For a network with n nodes and m weights, the sensitivity matrix has $O(nm)$ elements, each requiring $O(n)$ computations per element, giving $O(n^2m)$ calculations per step. For a fully recurrent network, this dominates the computational complexity and requires $O(n^4)$ computations per step. The algorithm works quite well on small networks, but the n^4 factor becomes overwhelming as the number of nodes increases.

The value of n in the $O(n^2m)$ equation is the number of recurrently connected units. Zipser's algorithm reduces this value by creating subgroups of nodes where sensitivity information is only passed between nodes of the same subgroup. All connections still exist in the forward dynamical system, the subgroups only affect the training of the network. Connections between subnets are treated as inputs. If g is the number of subgroups, then the speed-up of the sensitivity calculations is approximately g^2. For instance, dividing a network into two subsets ($g=2$) gives a 4-fold speed-up in computing the sensitivities. If the size of the subnets remains constant as the size of the network is increased (increasing the number of subnets), then the complexity of the RTRL algorithm is reduced from $O(n^2m)$ to $O(m)$.

The performance gains are substantial, but the question is whether the algorithm can train networks as well as the full RTRL. One might think that the subgrouping will limit the capabilities of the network to share nodes. This is not the case, however, since the activations of the network are unchanged -- it is still fully recurrent except in the training methodology. Even though the error propagation is limited to the subnets, all units have access to the activities of all other units, just not all of their sensitivities. Zipser's empirical tests indicate that they can solve many of the same problems, but for certain applications, networks trained with subgrouped RTRL require more PEs than when they are trained with full RTRL. In my experience, the subgrouping algorithm typically also requires more training epochs to reach the same MSE.

One caveat of the subgrouped RTRL training is that each subnet must have at least one unit for which a target exists since gradient information is not exchanged between groups. The problem can be solved by wrapping a feedforward network around the recursive network – creating a feedforward MLP with a fully recursive hidden layer. This is often termed a recurrent multilayer perceptron (RMLP) and is shown in Figure 3-16. The feedforward network is simply one additional layer that distributes the gradient between the groups. With simple extensions to the algorithms, multiple fully-recurrent layers can be added to the network.

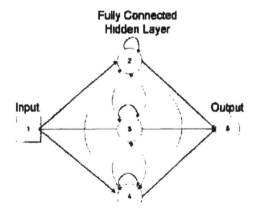

Figure 3-16: Diagram of a fully recurrent multi-layer perceptron (RMLP)

Dynamic Subgrouping with π

The goal of my method is to create local neighborhoods (subgroups) in the RTRL algorithm where the majority of the gradient information required for each node is confined to its local neighborhood. This requires organizing the recurrent PEs such that those that have strong temporal dependencies are neighbors. This technique replaces the static, preallocated grouping of Zipser's technique with a dynamic method of determining the best set of neighbors for each PE. This dynamic grouping provides faster and more robust training than Zipser's technique while maintaining its $O(n^2)$ performance.

First, the RTRL equations must be modified slightly to better suit the RMLP architecture described above. The time indices are now defined such that the input vector contains the external inputs from this time period plus the values of the PE outputs from the previous time period (i.e. the feedback).

$$IN_L(n+1) = [x(n+1), y(n)]$$

$$\pi_{kL}^{j}(n+1) = \varphi'(v_j(n+1))\left[\sum_{i\in B} w_{ji}\pi_{kL}^{i}(n) + \delta_{kj}IN_L(n+1)\right]$$

Next, we must determine the criteria we will use to group the PEs in the network. If we assume that each PE is responsible for updating the weights of the arcs which terminate at that PE (i.e. the incoming connections), then the PEs that have the highest sensitivities relative to those connections should be in the same neighborhood. For example, $PE\,j$ is responsible for updating all weights w_{jB} where B is the set of recurrent PEs. If we define

$$Z_{jk} = \sum_{L} \pi_{kL}^{j}$$

then the value of Z_{jk} provides a measure of how much $PE\ k$ affects the weights of $PE\ j$.

Thus the neighbors to $PE\ j$ should be the ones which have the highest Z_{jk}.

The "dynamic subgrouping with π" (DS-π) methodology implements the RTRL

algorithm with the subgroups chosen using the Z measure defined above. It should be

noted that since it requires the computation of the complete π matrix, this algorithm is no

more efficient than the full RTRL algorithm. It will, however, address how the

neighborhood technique with "optimal" switching will perform compared to the full

RTRL and Zipser's technique.

The test case for the DS-π algorithm is a function approximation problem where

we are trying to map a frequency doubler. The input to the network is a sinusoid with a

16-point period and the desired signal is a sinusoid with an 8-point period. This is a

nonlinear function since linear functions cannot "create" frequencies. The DS-π network

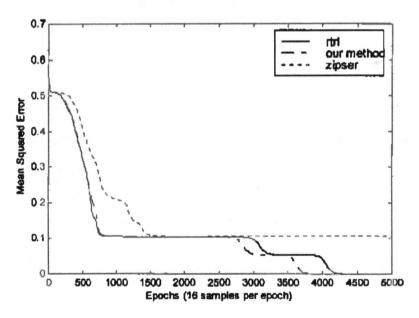

Figure 3-17: Average learning curve for the three algorithms using the frequency doubling problem

has 6 fully recurrent hidden layer PEs and one linear output node. Both Zipser's method and the DS-π method use two groups of three PEs. Each of the three algorithms was trained using the same five sets of random initial weights and the results were averaged to obtain the learning curves. Figure 3-17 shows the average learning curve for each algorithm. Notice that the full RTRL and the DS-π method performed nearly identically. In fact, in a few cases, DS-π actually trained in fewer epochs. The third set of initial weights led all three algorithms to a deep local minimum. The first 100 epochs mainly depict the learning curve from the other four initial conditions (notice that the DS-π method and RTRL are nearly identical here), whereas the last 400 iterations are dominated by the learning curve for the initial weights with the deep local minimum. Zipser's method performed worse on all 5 sets of initial conditions and couldn't solve the problem at all (even with more training) for the 3rd set.

In every application I have tested, the DS-π algorithm trains the networks in almost the same number of epochs as the full RTRL algorithm and performs significantly better than Zipser's subgrouping technique. The problem, however, is that the full π matrix is required to compute the neighborhoods. Since the computation of the π matrix is the computationally expensive part of the task, we have not gained anything here. This methodology, however, proves that the technique is feasible and that the all the gradient information is not necessary to train the networks. Also, when a neighborhood changes in the DS-π algorithm the gradient information from the ex-neighbor is discarded and new gradient information from the new neighbor starts building up. The technique using the full π matrix shows that this resetting and restarting of gradient information between nodes does not affect the performance of the algorithm. The DS-π algorithm will be used as an "ideal grouping" methodology since it uses all of the information of the sensitivities to determine the groupings.

Estimating the Z matrix

We now need an estimate of Z that will allow us to efficiently compute the temporal neighborhoods. The logical choice for an estimate of Z is to use the first-order estimate of the π matrix to compute Z. We start by writing out the equation for Z and simplifying:

$$Z_{jk} = \sum_L \pi_{kL}^j(n) = \sum_L \left(\varphi'(v_j(n)) \left[\sum_{r \in B} w_{jr} \pi_{kL}^r(n-1) + \delta_{kj} IN_L(n) \right] \right)$$

$$Z_{jk} = \varphi'(v_j(n)) \sum_L \left(\sum_{r \in B} w_{jr} \pi_{kL}^r(n-1) + \delta_{kj} IN_L(n) \right)$$

At this point, I will stop and discuss some grouping rules that I have implemented. First, unlike Zipser's work, the groupings do not need to be symmetric. *PE j* can be a neighbor of *PE k* without *PE j* being a neighbor of *PE k*. Thus, the baseline method is not a true grouping, but a linking of PEs which are sensitive to each other. A true grouping can be determined by modifying the grouping criteria to include both directions (e.g. $Z_{jk}+Z_{kj}$). Since symmetry is not being enforced, the methodology enforces the rule that *PE j* is always a neighbor of *PE j*. This does not have to be the case, but seems to be a reasonable assumption. Much of the gradient information from a recurrent network comes from the self-recurrent loop in each PE.

Since we assume that *PE j* is always a neighbor of *PE j*, we only need to compare the total sensitivity of all the other PEs. Thus, we do not need to worry about the $Z(j,j)$

terms which means that the δ_{kj} term can be removed. Reorganizing the summations slightly leads to:

$$Z_{jk} = \varphi'(v_j(n)) \sum_{i \in H} w_{ji} \sum_L \pi_{kL}^i(n-1) \text{ where } j \neq k$$

Expanding, we get:

$$Z_{jk} = \varphi'(v_j(n)) \sum_{i \in H} w_{ji} \sum_L \varphi'(v_i(n-1)) \left[\sum_m w_{im} \pi_{kL}^m(n-2) + \delta_{ik} IN_L(n-1) \right]$$

and

$$Z_{jk} = \varphi'(v_j(n)) \sum_{i \in H} w_{ji} \varphi'(v_i(n-1)) \left[\sum_L \left(\delta_{ik} IN_L(n-1) + \sum_m w_{im} \pi_{kL}^m(n-2) \right) \right]$$

Now we separate the equation into its first order parts (the direct contributions from the input vector -- when i=k) and the rest.

$$Z_{jk} = \varphi'(v_j(n)) w_{jk} \varphi'(v_k(n-1)) \sum_L IN_L(n-1)$$
$$+ \varphi'(v_j(n)) \sum_{i \in H, j \neq k} w_{ji} \varphi'(v_i(n-1)) \left[\sum_L \sum_m w_{im} \pi_{kL}^m(n-2) \right]$$

This is a very interesting equation. Let's say we approximate Z with the first order terms:

$$\widetilde{z}_{jk} = \varphi'(v_j(n)) w_{jk} \varphi'(v_k(n-1)) \sum_L IN_L(n-1)$$

Notice that the sum of the input scales all terms of Z_{jk} the same, so it can also be eliminated, leaving only:

$$\widetilde{Z}_{jk} = \varphi'(v_j(n)) w_{jk} \varphi'(v_k(n-1))$$

This is a very easy and computationally efficient method for estimating the Z matrix. It is conceptually appealing as well. You can see that this equation is a time correlation

93

between the derivatives of the non-linearities of the PEs j and k. If this were a static, linear network, the equation would simply be

$$\widetilde{z}_{jk} = y_j(n)x_j(n-1)$$

where y_j is the output of *PE j* and x_j is the input to *PE j* from *PE k*. This is a temporal version of Hebbian Learning. The first-order estimate of Z can be considered a nonlinear version of Hebbian learning. The derivative of the nonlinear function at the operating point determines the sensitivity of each PE to the current training input. Thus, we are calculating a correlation of the sensitivities of each PE. Which can also be considered as a correlation in the dual of the network.

If the PEs of the network use a tanh activation function, the estimate for the Z matrix can become a *local* rule. Local rules are advantageous because they are easily implemented in parallel and easily analyzed. Figure 3-18 shows the plot of $f'(net)$ versus $f'(out)$, which is equal to $f'(f(net))$, for a tanh PE. Since these two shapes are very similar,

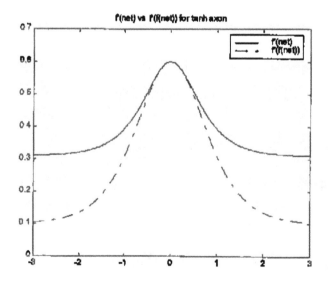

Figure 3-18: Using a tanh PE, f'(net) takes the same shape as f'(out)

the subgrouping criterion can replace $f'(net)$ with $f'(in)$ since the output of the source PE is the input to the destination PE. The new equation for Z becomes

$$\tilde{Z}_{jk} = \varphi'(v_j(n))w_{jk}\varphi'(y_k(n-1))$$

The local implementation assumes that each PE stores the weights for its incoming nodes locally and thus has access to all of its inputs and its output. This methodology was tested and produced identical results to the DS-FOE method.

Illustrative Example

The performance of the network trained with dynamic subgrouping and the first-order estimate of the π matrix (DS-FOE) will be compared against the full RTRL algorithm and Zipser's method. The example data set will again be the frequency doubling problem. Remember that for this problem, the DS-π algorithm achieves nearly

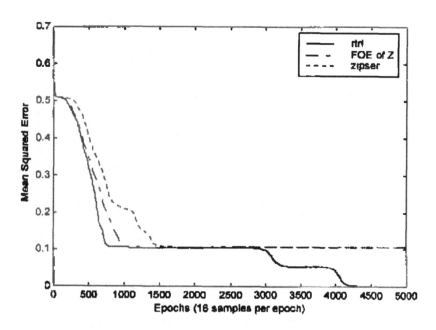

Figure 3-19: Average learning curve for the RTRL, DS-FOE, and Zipser's algorithms using the frequency doubling problem

identical performance to the full RTRL algorithm, so for clarity it is not shown in the plots. Figure 3-19 shows the learning curve averaged over 5 random initial conditions, as described previously. Notice that the DS-FOE method performs significantly better than Zipser's method yet not as well as the theoretical limit. The DS-FOE method could also not solve the frequency doubling problem on the third set of data in the allotted time. For the rest of the initial conditions (look at the first 200 samples of the learning curve), the full RTRL algorithm required 750 epochs to converge, the DS-FOE method required 1000 epochs to converge, and Zisper's method required 1500 epochs to converge.

Grouping Dynamics

One of the assumptions for the dynamic subgrouping algorithm is that the Z matrix, which represents the temporal "distance" between each PE, changes slowly. If the Z matrix changes rapidly, the neighborhoods will be constantly changing and a good approximation to the Z matrix will be impossible to achieve. Beyond that, the constantly changing neighborhoods will cause an increased number of transients in the training since each time a neighborhood changes, some of the gradient information is discarded. Figure 3-20 shows a plot of $Z(1,i)$ over time for the DS-π algorithm. This plot shows the distances from PE1 to each of the other PEs in the network. The three nearest neighbors of PE1 are shown at the bottom of the figure (remember that PE1 is always a neighbor of PE1). As expected, the curves are all smooth over the 100 epoch plot. Specifically, in the beginning of the training, the two PEs closest to PE1 are PE3 and PE4. After 15 epochs, PE 2 begins to have a larger influence on PE1 and replaces PE4 as a neighbor to PE1. Soon there after PE2 grows quickly and becomes the dominant neighbor to PE1. After approximately 80 epochs, PE3's influence on PE1 wanes and PE4 again becomes a neighbor of PE1.

This type of neighborhood plot is typical, and the smoothness and infrequent switching is common to all the applications I have tried. Occasionally, two PEs will have very similar values and will swap places frequently for a period of time. Normally when this happens, however, the Z values are fairly small and the gradient updates are being dominated by a much larger PE.

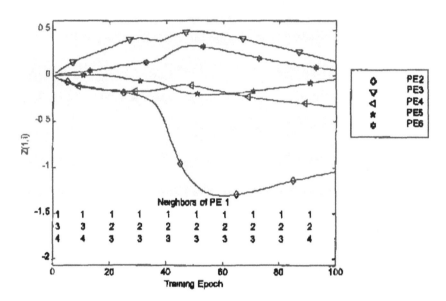

Figure 3-20: Plot of Z matrix over time showing the temporal neighbors of PE 1

Second Order Methods

Many researchers have found that first order gradient descent techniques like RTRL and BPTT do not provide satisfactory performance for training recurrent neural networks. Most commonly they use the extended Kalman Filter (EKF) (also known as the extended RLS algorithm). The EKF algorithm does not actually compute the dynamic gradients differently, it simply uses more information about the gradients to achieve a better estimate of the shape of the performance surface. Thus, the EKF algorithm still needs to use either BPTT or RTRL to compute the instantaneous estimates of the gradients. The dynamic subgrouping technique proposed here can be used to drastically speed up these computations in the EKF algorithm.

A more interesting application of the dynamic subgrouping mechanism may be in the actual EKF formulation. Just like the RTRL algorithm, the EKF algorithm is too computationally burdensome to be fully implemented. Instead, most typically use the decoupled extended Kalman filter (DEKF) approach that ignores all cross-terms between PEs in the covariance matrix. This simple approach is similar to Zipser's approach for RTRL. The dynamic subgrouping could be used at the DEKF level as well, allowing the network to extract enough data from the input to determine which portions of the covariance matrix are important. This is equivalent to determining which PEs in the system should be temporally clustered.

Summary of the Dynamic Subgrouping Algorithm

By using temporal self-organization in the RTRL algorithm we created a dynamic version of Zipser's method of subgrouped RTRL that can provide a robust $O(N^2)$ training method for recurrent neural networks. The method makes use of a first order estimate of the π matrix to compute a metric (Z) that is used to group the PEs in the network. This method performs roughly halfway between the performance of Zipser's method and that of the "optimal" grouping. The current method of estimating the Z matrix is not only first order but also has a memory of only one step. It seems reasonable to believe that other estimates of Z may provide better performance without significantly increasing the computational complexity beyond $O(N^2)$.

The most interesting theoretical result is that with an "optimal" determination of the subgroups, the dynamic subgrouping algorithm achieves nearly identical performance as the full RTRL algorithm. In fact, it will occasionally outperform it. This could be because the PEs that contribute the least to the gradients of certain weights are not necessary and may only contribute noise to the overall gradient. By choosing only the PEs that contribute the largest fraction of the gradient, the DS algorithm may be achieving a PCA-like affect of removing the noise in the gradient.

CHAPTER 4
APPLICATIONS AND RESULTS

The previous chapter presented the theory and background information required for each of the three architectures I have created by applying temporal locality and activity diffusion. For illustration, each section also included a simple application for each architecture. This chapter presents the results of applying these architectures to a series of practical and/or more extensive problems. Again, we will begin with the SOTPAR architecture that is based on the self-organizing map, then the SOTPAR2 architecture based on the neural gas algorithm, and lastly the dynamic subgrouing algorithm based on real-time recurrent learning (RTRL).

SOTPAR

The SOTPAR architecture uses activity diffusion to create a spatio-temporal self-organizing map. The SOM has been used for many different applications, but in its most simple form is a vector quantization technique. The first application uses the SOTPAR dynamics to create a landmark discrimination and recognition mechanism. The complete SOTPAR architecture could be trained to solve this problem, but for improved speed of training, we modified the architecture to allow one-shot training. The second application of the SOTPAR architecture is based on the work of Ruwisch, et.al. This application uses the SOTPAR to organize a linear SOM for sequences of spoken phonemes. When the network is repetively trained with phoneme sequences, it organizes itself such that the phonemes are sequentially located in the map.

Landmark Discrimination and Recognition for Robotics

Pedro Kulzer, from the E.E. robotics group, and I have applied the SOTPAR to a problem of determining the shape of a landmark using only local information obtained from a very simple robot. [Eul96b] The robot has only forward and lateral infrared sensors that can detect if it is near an object. There is no long distance vision or perspective obtained by these sensors. The robot uses wall-following techniques (similar to a blind person who can only feel the walls of a building) to determine the shape of the landmark. Using the limited information obtained by the shaft encoder and object detection sensors, the robot must discriminate between different landmark shapes while it traverses the walls of an object/landmark. Much of the problem description and robot control algorithms and some of the figures come from Pedro Kulzer's Master's Thesis [Kul96].

The robot wanders about its "space" until it finds an object in front of it. The robot may approach this object from any possible direction and orientation. In order to simplify the procedure, when the robot detects an object, it first *docks* with the object. Docking consists of rotating the robot until its right side is parallel to the landmark's wall. The robot is determined to be parallel to the landmark when the front-right and rear-right sensors indicate equal distances to the landmark. After successful docking, the robot follows the walls of the landmark in a clockwise direction. This is the simplest algorithm

101

for circumventing an object, but creates difficulties for the processing of the data. Figure 4-1 shows the docking behavior and a simplified wall-following diagram.

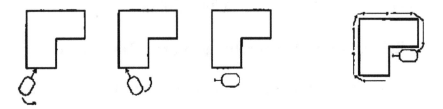

Figure 4-1: Docking behavior and an example of wall-following [Kul96]

One of the difficulties with the wall-following approach is that the data collected by this simple robot is imprecise. The wheels and gears have a tendency to slip and the turning data and segment distances are often incorrect. Additionally, the wall-following control algorithm can overshoot or undershoot the turns, just like any other control algorithm. To obtain more accurate information, the speed that the robot circumvents the object must be much slower. The goal of our approach is to allow the robot to operate at higher speeds by creating an algorithm that will accept less precise local turn data. Figure 4-2 shows examples of turns at normal speed, slow speed, and faster speed.

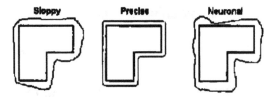

Figure 4-2: Robot wall-following at different speeds [Kul96]

The main difficulties with the landmark discrimination problem are:

• the exact location where the robot first encounters the landmark is unknown

- the circumference of the object is unknown (how do you know when you have traversed it completely)

- the robot would like to learn the landmarks very quickly, preferably in a single circumvention

- the slipping of wheels and gears creates noisy and inexact data during turns

- symmetric shapes will create uncertainty in the exact location of the robot relative to the landmark (e.g. a square has 4 identical sides and angles)

SOTPAR solution

To simplify the problem we added a simple compass to the robot. This helps solve the symmetry problem since now we not only have turn angles and distances, we have orientation information. At preselected intervals the robot collects its turn angle and compass settings. The relative coordinate turn angle is the derivative of the compass setting and thus this creates a dynamic state-space description of the motion of the robot. If a compass is not present, the turn angle can be derived from the wheel motions and the compass would be the integral of the turns, but the compass would drift over time and would soon be unusable.

The sampled data can be constructed into a trajectory through turn-angle/compass space. The SOTPAR can be used to map such trajectories. Figure 4-3 shows the trajectory the robot must follow to traverse the L shaped landmark. The trajectory begins when the robot is heading north ($0°$) after making a right turn at the bottom-left portion of the landmark. The interpolated turn slowly moves back to zero and then back to $-\pi/2$ as the robot turns right again and begins to head east. The trajectory is very difficult to map

103

because there is an overlap in the input space where the trajectory doubles back on itself.

This corresponds to the concave corner in the figure and is shown as the line moving

from approximately $(3\pi/2,-\pi/2) \rightarrow (\pi,\pi/2)$ and vice-versa. The trajectory passes through

the same points in state space and the only difference is the direction of travel. Without

temporal information, the standard SOM can not properly map this trajectory.

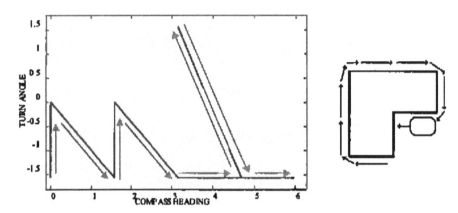

Figure 4-3: Robot trajectory around the L-shaped landmark

A SOTPAR network with 6 strings and 10 PEs was created to map the target

trajectory. Since this is a complicated target trajectory that is 40 samples long, training a

single long SOTPAR network for this problem is very difficult. Multiple strings of PEs

provides improved flexibility in the mapping of complex trajectories by allowing

different clusters of PEs to move independently. Each string of PEs is like a single 1D

SOTPAR network. The only mechanism linking the 6 strings of PEs is the wavefront of

enhancement that travels from the last PE of one string to the first PE of the next. The

neighborhood training of the strings is local and does not cross string borders.

The SOTPAR was trained with noisy versions of the target trajectory (the turns

and compass headings) intermixed with noise that represents random searching motions of the robot looking for landmarks. There was no supervision or indication as to which part of the input was noise and which was signal. Table 4-1 shows the training parameters for the SOTPAR network. The learning rate and SOM neighborhood width are generic parameters that are typically annealed from high to low values during training. The annealing of the learning rate allows for fast but coarse movements of the PEs at the beginning of training and slower fine-tuning of the PE positions at the end of training. The annealing of the neighborhood width organizes the network globally at first and then fine tunes the local neighborhoods at the end of training. The two new SOTPAR parameters are the spatio-temporal parameter, β, and the temporal decay parameter, μ. The spatio-temporal parameter is annealed similar to the other SOM parameters and allows the network to concentrate on temporal ordering first (high β) and then on fine tuning the spatial map at the end of training. The temporal decay parameter sets the feedback proportion of the leaky integrator at each PE. I typically use a value of 0.7 which provides a good tradeoff between forcing the network to always move forward and allowing repeated PE firings.

Table 4-1: Training parameters for robot trajectory

Training Parameter	Value
Spatio-temporal parameter, β	Linearly annealed from 2 to 0.3
Learning rate, η	Linearly annealed from 0.01 to 0.002
SOM neighborhood width	Linearly annealed from 2.5 to 0
Temporal decay parameter, μ	0.7
Number of training epochs	100
Number of strings	6
Number of nodes per string	10

Figure 4-4 shows the trained PE weights of each string mapped back to the input space and represented as X's (O's represent the first PE in each string). Four (labels 1-4) of the strings mapped to different locations in the trajectory (as labeled on the right side of the figure) and two others (labels 5 & 6) account for most of the inter-signal noise. As the robot moves along the periphery of the object, the nodes of the SOTPAR will fire sequentially.

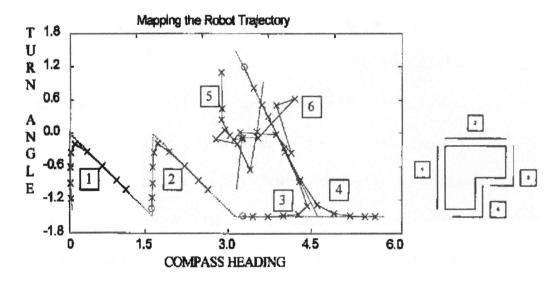

Figure 4-4: SOTPAR mapping of L-shaped landmark trajectory

The SOTPAR dynamics use temporal information to smooth noisy signals and can also gracefully handle time-warping. If the input data is noisy, the temporal wavefronts that move through the network will influence the selection of the winning PEs and be able to ignore a certain amount of noise. The spatio-temporal parameter, β, determines how much noise the system will accept. If β is set to 0.5, then the temporal information in the network can influence the network to choose a PE that is up to 0.5 units farther away from the spatial input than the closest spatial PE. Thus, β should be set based upon the dynamic range of the input and the amount of expected noise in the signal. Of course increasing β also decreases the ability of the network to discriminate between two similar but different temporal patterns.

Figure 4-5 shows two trajectories with 0.1 amplitude and 0.3 amplitude zero-mean additive noise. The trajectories are significantly different than the training signal, but the dominant temporal pattern is still clearly visible. Figure 4-6 shows the sequence of winning PEs and the SOTPAR enhancement in the network for both trajectories, with random noise interspersed. Remember that a diagonal line in the plot of winning PEs means that the PEs fired in a sequential order, as desired. β was set to 0.5 and thus the network should be able to use the temporal information in the signal to remove much of the variation in the signal. The trajectory with 0.1 amplitude noise is shown in the first 40 points of the figures. Notice the straight diagonal line that shows that the PEs fired in perfect sequential order. The wavefront is perfectly synchronized with the signal and moves along just ahead of the previous winning PE, helping the network choose the

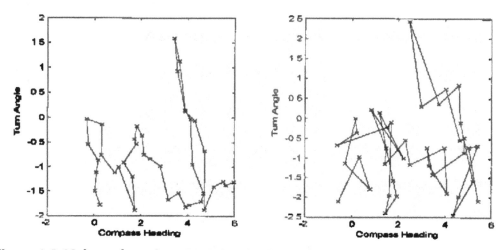

Figure 4-5: Noisy trajectories with 0.1 and 0.3 amplitude noise

107

correct PE. The second trajectory is shown between points 60 and 100. The SOTPAR was still able to correctly map the majority of the signal and only three times miss classified a point on the trajectory. These locations are clearly shown on the plot of winning PEs and also show up as a "dimming" of the enhancement wavefront in the enhancement chart.

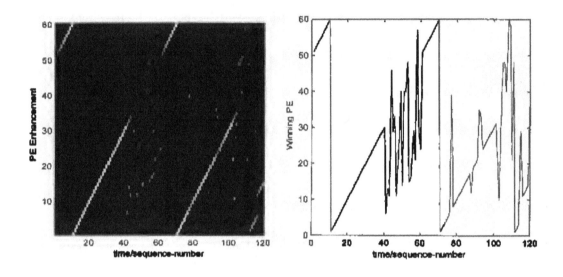

Figure 4-6: Enhancement and winning PEs for noisy trajectories with +/- 0.1 and +/-0.3 amplitude noise

If the input data is time-warped, the memory kernel in the SOTPAR allows the network to either skip PEs for shorter patterns or fire PEs more than once for longer patterns, without greatly reducing the wavefront strength of the network. Maintaining the wavefront strength allows the network to continuously smooth spatial noise with temporal information. Two sequences were created that warped the 40-point trajectory to 56 points and 30 points by upsampling and downsampling the signal. Figure 4-7 shows the enhancement of each PE over time and the sequence of winning PEs. The longer trajectory is shown in the first 56 samples and the enhancement plot shows that the wavefronts periodically die out and restart at the same PE one time period later. The plot of the winning PEs also shows that certain PEs are fired twice so that the SOTPAR can warp the signal back onto its output map. Samples 72-102 show that the network periodically skips a PE to adjust the network to the faster sampling of the shorter trajectory. The enhancement shows disconnected wavefronts that continue in the network just ahead of where the previous one ended.

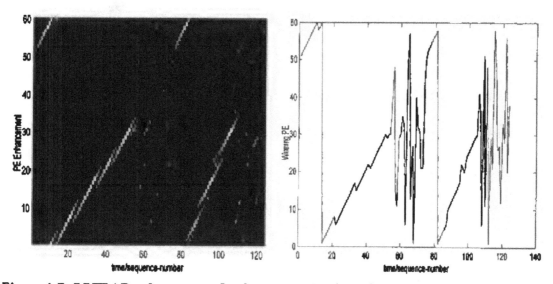

Figure 4-7: SOTPAR enhancement for time-warped trajectories

The ability of the SOTPAR network to remove noise and deal with time-warping is very adventageous, but even the relatively short 100 epoch training time required for this network is too long for the robotic application. As a general rule, increasing the constraints of the system by using application specific information leads to simplified and faster training. By adapting the SOTPAR to the specific characteristics of the landmark object recognition problem, we created a new network that can be trained much faster. First, since the robot can detect when it encounters an object, multiple simplifications can be made. Each object will be represented by a single string of nodes, each with all the

SOTPAR dynamics as defined above. Additionally, since the exact location where the

robot first encounters the object is unknown, the endpoint of the string will wrap around

to the starting point to form a ring. Thus, a wavefront will be created and rotate around

the ring from any starting position. A one shot training algorithm is implemented which

trains the landmark string as the robot circumvents the object for the first time. After the

robot docks with the object, it follows the walls of the landmark in a clockwise direction.

At predefined intervals, a PE is added to the current string, with its weights set to the

current turn angle and compass heading. This continues until the robot completely

circumvents the landmark at which point the end of the string is connected to its

beginning, forming a ring network.

Another modification designed by Pedro Kulzer [Kul96] makes the network more

forgiving to imprecise turn angles and also provides an interesting link to biology and

other technologies. The input of the network is first pre-processed by a set of PEs based

upon a "tuning curve". The tuning curves shown in Figure 4-8 provide a more gentle roll

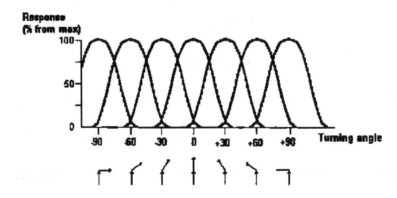

Figure 4-8: Tuning curve for turn angles [Kul96]

off of the activation of a PE with increased distance between the stored angle and the actual angle. The stored angles are discretized such that they are stored in increments of 30 degrees. Each node responds to one turn angle based upon the "tuning curve" shown in the figure.

This tuning curve mechanism allows for an interesting rationale for the one-shot training algorithm. The tuning curve acts as a preprocessor that resembles a radial basis function network or a fuzzy logic system. In biology, SOM-like structures are commonly found doing such tasks, clustering sensory inputs. Each preprocessor PE represents a certain turning angle and fires when the current turn matches its stored turn. The one-shot training algorithm can be mimicked if each tuning curve PE is linked to the SOTPAR chain and the weights are trained in a Hebbian manner with very high learning rates, as shown in Figure 4-9. The figure shows only one set of weights from the preprocessor to PE 4 with the weight between 0 degrees and PE 4 being much stronger than the others.

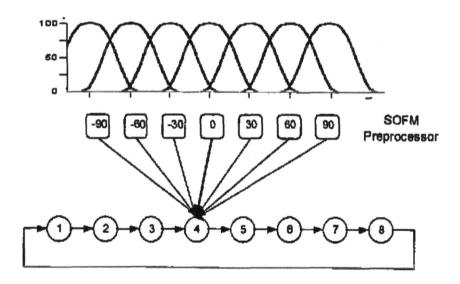

Figure 4-9: Preprocessor feeds SOTPAR string using fast Hebbian learning

The SOTPAR dynamics are used for the recognition of the objects. The turn angles and compass headings are fed into the network and wavefronts are created as the robot traverses the outside of the landmark. A new ring is used for each landmark. Three different properties found in the SOTPAR provide useful information in the landmark recognition problem.

- *Landmark Recognition.* As the robot moves around an unknown landmark, the network with the largest continuous wavefront will determine which landmark the robot is processing. To demonstrate this capability, I have simulated a circumvention of two different landmarks shown in Figure 4-10, the L-shaped landmark and the square shaped landmark. These two landmarks are difficult to discriminate since the turns that are required to traverse these two landmarks are very similar, differing only at two locations. Zero-mean random noise is added to the inputs to simulate the "sloppy" turning and wall-following algorithm that the robot will use. After the one-shot training on both landmarks, a new noisy input set was created for the L-Shaped landmark and run through both rings from an *unknown starting point*. The maximum activity of the two strings of 8 PEs is shown in Figure 4-10. The figure shows that the activity wavefront moves through the L-shaped landmark ring with little attenuation. The square landmark ring, however, starts to build up as it traverses the left and top of the figure and then decays rapidly as two wrong turns are made.

- *Endpoint Detection.* Endpoint detection can be automatically determined while the network is training. When the robot passes the point where it first encountered the landmark, a second wavefront will begin forming at the beginning of the network and will move around the ring. The distance between the current location (the position of the original wavefront which should always be at the most recently added node) and the redundant wavefronts is the estimated circumference of the landmark. Figure 4-11 shows the wavefronts moving around a network being created as the robot traverses the L-shaped landmark. When the robot detects a sustained secondary wavefront, it should clip the network at the point where the secondary wavefront begins and reconnect the tail to the start of the ring.

- *Location Relative to the Landmark.* As the robot traverses the landmark, the location of the wavefront tells the robot its location relative to the landmark. This is not absolute information, it only describes where the robot is relative to the first time it traversed this landmark. Absolute location information, however, is not necessary. Figure 4-11 shows the wavefronts from a sample traversal. The current winning PE provides the best estimate as to the location of the robot relative to the landmark.

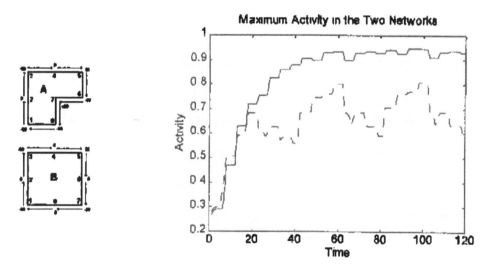

Figure 4-10 - Maximum Activity in the Two Networks (Solid line = L-shape, Dashed line = Square)

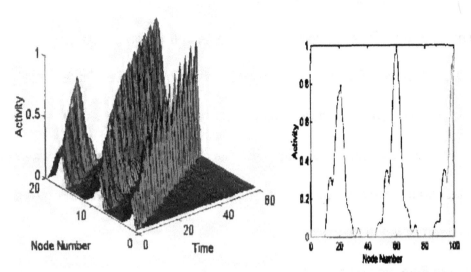

Figure 4-11 - The wavefronts moving through time and space tell the circumference of the landmark

Real data collected from the robot

The SOTPAR methodology described thus far has been based on the addition of

the enhancement and distance measure. Another possibility, as discussed in Goppert

[Gop94] and as implemented by Pedro Kulzer, is to multiply the enhancement and the

distance measure. The multiplication lowers the noise resistance of the network since a

single bad input will bring the output of the network down to zero, but it improves the

discrimination between two similar landmarks. Kulzer implemented both the additive

and multiplicative models and achieved very similar results. The results presented in this

section are based on the multiplicative model. Additionally, the endpoint detection

algorithm described above was not discoverd until after the tests with the real robot were

completed. This adversely affected the results since the proper endpoints could not

always be obtained. One further difference between the theoretical results and the actual collected data is that the existing robot did not have a compass input. Therefore, the discrimination is much more difficult using only local turn information.

The data was collected from a prototype robot as described previously. The first landmark that was traversed is a rectangular sofa. Without the compass data, the turning information and stopping points were very noisy. Figure 4-12 shows a diagram of the sofa and a representation of 13 traversals around the sofa by the robot. The traversals started at random locations and the endpoint was determined by trying to determine when the robot had gone through 360 degrees. The figure plainly shows how noisy and inaccurate both the turn data and endpoints were with this robot.

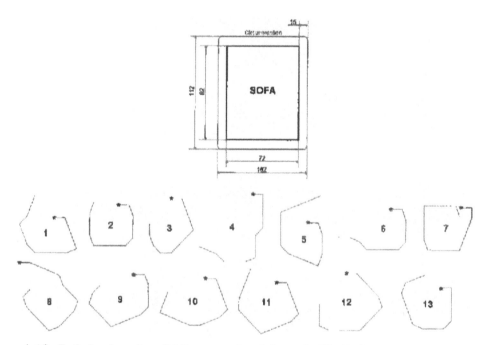

Figure 4-12: Sofa landmark and 13 traversals of the sofa [Kul96]

Similarly, a sofa and footstool combination was circumvented. This combination represents an L-shape. Figure 4-13 shows a diagram of the landmark and six traversals of the landmark. Notice how difficult it will be to discriminate between the sofa and the sofa/chair landmarks using the collected data. The two sets are very similar and the chair

portion of the landmark is noticeable but not a dominant feature of the datasets. For

additional testing, a non-polygonal landmark was traversed as shown in Figure 4-14.

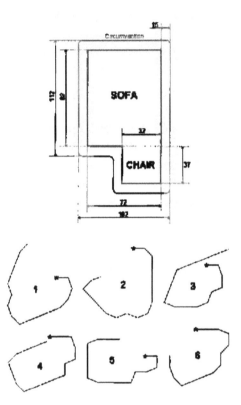

Figure 4-13: Sofa and chair combination landmark [Kul96]

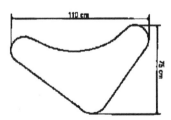

Figure 4-14: Non-polygonal landmark [Kul96]

With 3 landmarks and multiple circumventions for each landmark, there are many

combinations of tests that could be done. Each circumvention can be used to create a network which can then be tested against other networks and other circumventions. First, a comparison of the sofa and sofa/chair landmarks was done. A circumvention from each landmark was randomly selected and used for testing networks which were created from all other circumventions. The testing resulted in an 18.8% error rate, much of which can be explained by improper stopping points. Table 4-2 contains a sampling of the discrimination tests and presents the results by showing the ratio of the output of the correct network to the output of the incorrect network. A value of 1 or greater shows correct discrimination. The values in the table that are greater than one are bolded and show incorrect discrimination.

Table 4-2: Sofa vs. sofa/chair landmark discrimination ratios [Kul96]

	test1	test2	test3	test4	test5	test6	test7	test8
Sofa	1.51	1.65	0.98	1.13	0.90	1.23	1.61	1.99
Sofa/Chair	1.34	1.37	1.40	1.32	1.22	0.90	1.35	1.20

Next, the non-polygonal landmark was compared against both the sofa and sofa/chair landmarks. The error rate here was 12.5% which implies that the non-polygonal landmark is easier to discriminate. Table 4-3 and Table 4-4 show a subset of the data collected, again using the ratio of the correct versus incorrect network outputs to determine the quality of the results.

Table 4-3: Sofa vs. non-polygonal landmark discrimination ratios [Kul96]

	test1	test2	test3	test4
Sofa	2.78	2.50	0.97	0.86
Non-Polygonal	1.05	1.33	1.40	1.77

Table 4-4: Sofa/chair vs. non-polygonal landmark discrimination ratios [Kul96]

	test1	test2	test3	test4
Sofa/Chair	1.37	2.18	1.51	1.42
Non-Polygonal	1.30	1.25	1.30	1.39

Summary

The typical approach to solving the object recognition problem in the Robotics group at the University of Florida is to store information about each of the corners of the object and the distances between them. Then with AI techniques they would attempt to recognize the object. This algorithm, however, works only for polygonal landmarks and is very brittle when the robot generates imprecise turning data.

The SOTPAR methodology described above performs better than the techniques being used at the time. Although the results are preliminary, our architecture has many advantages over the typical approach, including:

- The landmark can be of any shape, not only polygonal

- Space-warping (time-warping in terms of the neural network) created by varying speeds of the robot is gracefully handled

- The matching process is straightforward and resistant to noise, and does not require excessive amounts of computation

- The position in the circumvention is easily determined

- Endpoint detection can be done automatically

- Rotations and translations of the landmark do not affect the performance [Kul96]

Self-Organization of Phoneme Sequences

In 1993, Ruwisch et. al. [Ruw93] created parallel hardware versions of SOM

networks based on the reaction-diffusion equations. The fundamental concept is that

each PE is part of an active medium and when one PE fires, it eventually causes its

neighbors to fire as well. Once a PE fires, it remains active until a global reset is received

from the controller. When a sample input is presented to the network, the output of each

PE is proportional to the closeness of the match between the PE's weights and the input.

A global threshold is then progressively lowered until a PE reaches saturation and fires.

The PE that fires then triggers its neighbors as its activity diffuses to neighboring PEs.

This causes a wavefront that travels through the network. The wavefront moves at a

velocity c and at its simplest can be represented by the equation

$$\Phi(r,t) = H\left(ct - \left\| r_{win} - r \right\| \right)$$

where H is the Heaviside (step) function, r is the position of a PE in the lattice, and r_{win} is

the position of the winning PE. Although this work uses a moving wavefront through

time and space, they use it solely to implement the SOM algorithm in parallel. They do

not use it to create a spatio-temporal mapping of the input.

Although the equation bases the wavefront on the position of the winner, it is not

necessary to actually compute the location of the winner; the active medium does this

automatically. They use this wavefront to train the network. When a PE becomes active,

it can begin to train its weights (i.e. move them toward the input). Since the PEs which

were closer to the winner will be active for a longer period of time, this method naturally

implements the neighborhood function of the SOM in parallel. The wavefront

119

propagation and training continues until a global reset is received from the controller and

the next input sample is presented to the network.

More recently, they expanded their active medium concept by including a scaled

down version of our SOTPAR dynamics in their network. They modified and simplified

the SOTPAR dynamics such that it could easily be implemented in hardware. The

equation for the temporal enhancement wavefront is:

$$\Psi_t(r,t) = H\big[ct_i - d_{win}(r)\big]H\big[d_{win}(r) - ct_i + b\big]h(r,t)$$

where H is again the Heaviside function and $d_{win}(r)$ is the distance to the winning PE.

They also include a "history function", h(r,t), which precludes a PE from firing more than

once during the presentation of a single sequence. This history function keeps the

network wavefronts from reversing direction and retraversing a portion of the network.

The two Heaviside functions create a wavefront that is in the form of an expanding

concentric wave crest with a specified width b. Figure 4-15 shows an example of the

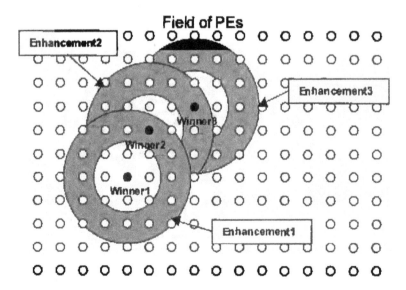

Figure 4-15: Ruwisch enhancement through a 2-D SOM [Ruw97]

wave crests moving through a 2-D SOM including the history function and successive winners. As in the SOTPAR, the winner is chosen as a combination of the spatial and temporal distances:

$$r_i^{wm} = \arg\min\left(\|U_i - W_r\| - \beta\Psi_{i-1}(r,t_i)\right)$$

The wavefront spreads until a global reset is sent from the controller. Notice that the only information used in the network is from one step in the past, other than the history function that does not provide significant information about the past. The attenuation and reinforcement of waves in the SOTPAR was not used.

The elegant portion of this methodology is that the process used to train the network (implement the neighborhood training in the SOM) and the process used to add the temporal information to the network (the limited SOTPAR dynamics) can use the same wavefront. With one activation of the active medium, both the temporal information and the training information can be spread throughout the network in a fully parallel manner. As the wavefront spreads, the PEs are trained both spatially and temporally.

I have recreated their methodology and tested the system on a set of data similar to that reported in [Ruw97]. The network is a one-dimensional string of 36 PEs. The input is phonemes from the TIMIT database of labeled and segmented speech. The subdivision of each word into phonemes was done by a TIMIT computer algorithm that sometimes produces less than ideal results. The acoustic signal was sampled at 16KHz and filtered by a set of bandpass filters. For simplicity, we used only three filters whose passbands were 0.6-1.0 Khz, 1.0-3.5 KHz, and 3.5-7.4 KHz. Each phoneme was averaged over the entire duration to produce a single feature vector per phoneme. The two words used were "Suit" and "Small". Suit has three phonemes ('s','oo','t') and small has 4 phonemes ('s','m','a','l').

A SOTPAR string of 36 PEs was then trained using the wavefront methodology described above. The history function and enhancement were reset after the complete presentation of each word. Table 4-5 shows the parameters for the training of the Ruwisch dynamics. The spatio-temporal parameter, learning rate, and SOM neighborhood width are all exponentially decayed. Often times an exponential decay improves the training time of the network over a linear decay of the parameters. The wavefront width and speed determine the number of PEs used to store each phoneme. In this case, the wavefront moved 4 PEs per sample of training data and was 3 PEs wide, providing 3 PEs to map the spatial characteristics of each phoneme. The 36-PE network trained for 200 epochs over the training set of 6 instances of each word. For a training set this size, using a wavefront more than 3 PEs wide to store each phoneme is excessive. For example, a 50-PE network with a wavefront speed of 7 and wavefront width of 5, did not train as well as the above network. There was enough flexibility in the network to map multiple phonemes to each phoneme neighborhood. Three of the 6 instances of the word suit moved in the same direction as the 6 instances of the word small.

Figure 4-16 shows the probability of a phoneme firing each PE (i.e. being vector quantized to each cluster) using the Ruwisch dynamics. For example, for the 's' phoneme in the top left-hand plot, PEs 9 through 14 fire exclusively. *The dynamics of the system organized the rest of the phonemes such that temporally neighboring phonemes will be located in spatially neighboring regions.* If you follow the plots from top to bottom, the phonemes from suit are stored in the network from left to right and the phonemes for the

Table 4-5: Training parameter for phoneme mapping

Training Parameter	Value
Spatio-temporal parameter, β	Exponentially decayed from 1 to 0.3
Learning rate, η	Exponentially decayed from 0.03 to 0.001
SOM neighborhood width	Exponentially decayed from 8 to 1
Number of training epochs	200
Number of Nodes	36
Wavefront speed	4 PEs per sample
Wavefront width	3 PEs

word small are stored in the network from right to left. The figure shows a few locations

where incorrect PEs fire, this can usually be explained by the poor segmentation of the

phonemes. For instance, the 'm' phoneme has a small peak in the 'u' region. This is due to

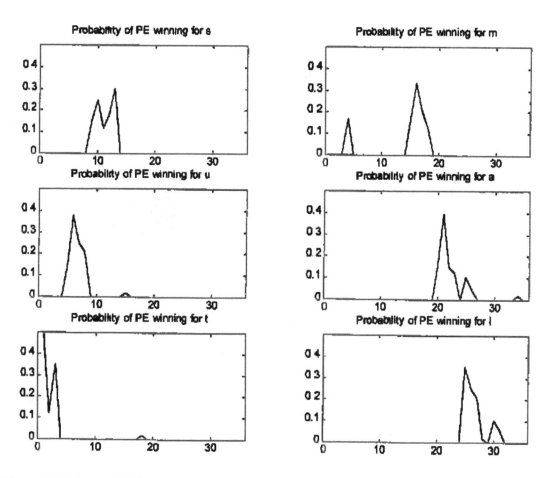

Figure 4-16: Probabilities of each PE firing for each phoneme

the improper segmentation of the 'm' phoneme that included some of the 'a' phoneme that follows it. Since the 'm' is much lower energy than the 'a', the small amount of 'a' contained in the 'm' created enough higher frequency energy to make it look like the 'u'.

The key attribute of using the traveling wavecrest version of the SOTPAR for this application is that *the SOTPAR dynamics allow us to use multiple PEs to store spatial variations (corresponding to differences in the input) and yet still create global temporal ordering between these local clusters.* Figure 4-17 shows an intensity coded image of the reference vectors for each PE in the network. It is clear from a cursory look at the image that the reference vectors are clustered into regions approximately 3 to 5 PEs wide. Since the winning PEs vary from speaker to speaker, the actual phonemes are clustered in a neighborhood of approximately 3-5 PEs in the output map, even though the wavecrest was only 3 PEs wide.A good example of how the network captured the spatial variations

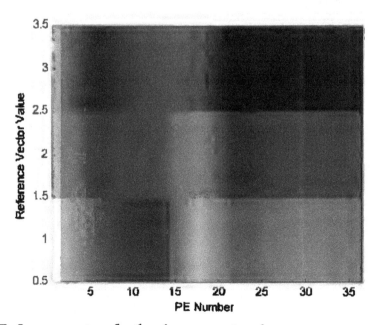

Figure 4-17: Reference vectors for the phoneme network

of each phoneme in the network is the region between PE 7 and PE 20. PEs 7 through 14 make up the 's' phoneme showing most of the energy in the upper frequency bands. PEs 15-20 represent the 'm' phoneme showing most of the energy in the low frequency bands. Notice how from left to right there is a lot of variation in the upper frequency bands of the 'm' phoneme. This is caused by the poor segmentation of the TIMIT database as the 's' sound is leaking into the 'm' phoneme. The SOTPAR network, however, is mapping and clustering these variations properly.

Figure 4-18 shows a graph that intensity codes the enhancement and winning PEs over time. The PEs receiving temporal enhancement are shown in shades of gray, with brighter indicating more enhancement, and the winning PE is shown in white. In the Ruwisch dynamics, however, all enhancement is at the same level, so there is only one shade of gray in the graph. On the horizontal axis is time, where each phoneme is presented for each word repeatedly. The first column represents the 's' of the first version of the word 'suit'. The PEs are on the vertical axis. Thus, the winning PE for the 's' of the first version of the word 'suit' is PE 13. According to the dynamics of the system, the winning PE causes the enhancement wavefront to travel in both directions from the winning PE. In this case, the wavecrest is 3 PEs wide and travels at a speed such that it is 2 PEs from the last winning PE. Thus, PEs 8-10 and 16-18 receive an enhanced chance of winning the next competition. The next winner is PE 8 which then creates two more wavefronts. Notice that in these dynamics, there is no building of energy from winner to winner, only the last winner determines the enhancement. The amount of enhancement is a constant determined by the wave equations. Notice in the graph that the 3-phoneme sequences (suit) move in the upward direction (from higher to lower numbered PEs) and the 4-phoneme sequences (small) move in the downward direction (from lower to higher

numbered PEs). This is consistent with the probability plots shown in the previous

figure.

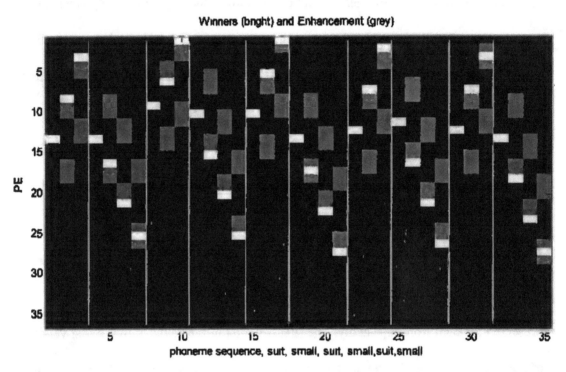

Figure 4-18: Winners and enhancement for the Ruwisch dynamics for the words 'suit' and 'small'

To verify that the temporal dynamics were responsible for this organization, I set

the spatio-temporal parameter, β, to zero. All other training parameters are identical to the

parameters used for the previous network. This short-circuits all temporal information in

the system and defaults to the static SOM. The probability plots are shown in Figure

4-19. Notice that there is no temporal organization whatsoever in the network and that a

few of the phonemes were trained into multiple spatial locations (e.g. the 'u' phoneme).

This confirms that the temporal dynamics of the SOTPAR are responsible for the spatio-

126

temporal organization of these maps. A comparison of this figure and Figure 4-16 shows

how the temporal information significantly cleaned the output of the network.

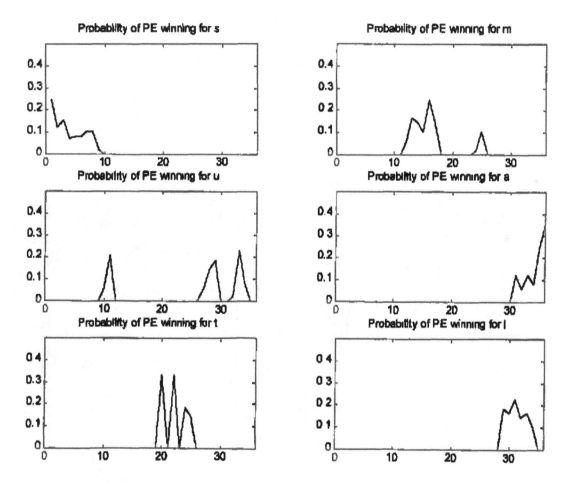

Figure 4-19: Probabilities for each PE without using the SOTPAR dynamics

As a final experiment, I tested the use of the full SOTPAR dynamics in this

application. The history function was retained to avoid backtracking in a one-dimensional

map. The main difference between the full SOTPAR dynamics and the partial SOTPAR

dynamics discussed above is that the full SOTPAR dynamics include the strengthening of

the enhancement when multiple correct firings occur sequentially. The more correct

sequential information the network receives, the more it becomes convinced it is recognizing a known pattern and the higher is the enhancement for the next phoneme. For the first two phonemes of each word, there should be little difference. For the last phonemes of each word there should be less noise with the full SOTPAR dynamics. Figure 4-20 shows the winners and enhancements using the full dynamics. Notice that the enhancement grows brighter after successive correct firings. Also, the wavefronts that move off in the wrong direction eventually die away.

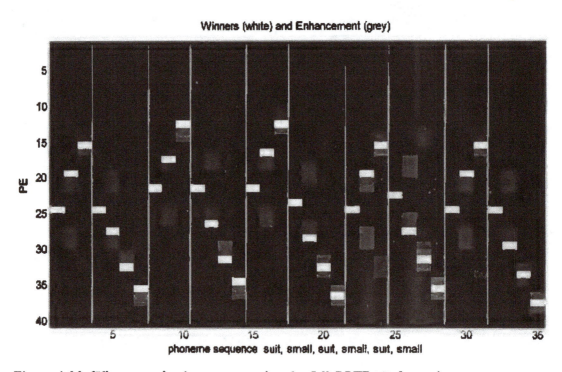

Figure 4-20: Winners and enhancement using the full SOTPAR dynamics

Figure 4-21 shows the probabilities for each phoneme to fire each PE. The results are very similar to the partial SOTPAR dynamics because the sequences are very short. The short duration does not allow the enhancement to grow significantly higher than the simplified one-step enhancement. The final phonemes are, however, more localized in the full SOTPAR dynamics than in the partial dynamics. One interesting aspect of using the

128

full SOTPAR dynamics is that the network is more tolerant of using large wavecrests. The Ruwisch dynamics could not properly map the test data with a 5 PE wavecrest. The full SOTPAR dynamics, however, provided enough extra information in the network for it to properly map all of the 12 words in the training set.

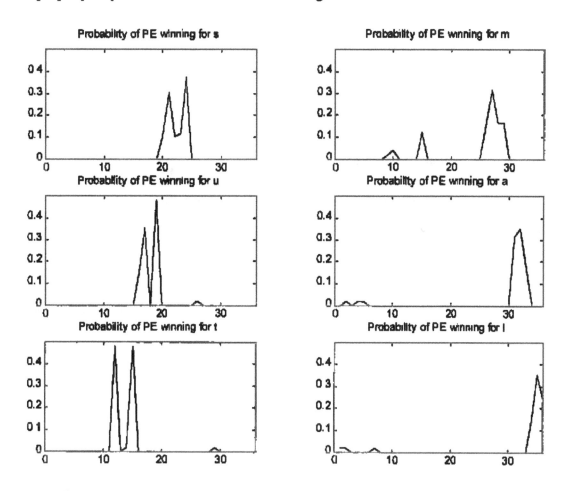

Figure 4-21: Probabilities for each PE winning using the full SOTPAR dynamics

Summary

This application showed that the SOTPAR dynamics can create a flexible architecture that allows phonemes to be stored among a cluster of PEs for local spatial flexiblity and yet these local groups can be globally organized with temporal information.

Each cluster represents a single noisy phoneme and adjacent clusters represent temporally sequential phonemes in the training set. This organization is created by waves of activity moving throughout the network. The temporal dynamics help remove some of the variability in the vector quantization produced by the network and the temporal ordering can be used to reduce the interconnectivity of neural networks designed for word recognition.

SOTPAR2

This section illustrates two applications of the SOTPAR2 network. The SOTPAR2 network uses temporal Hebbian learning and activity diffusion to determine temporal neighborhoods for each PE in a neural gas architecture. The first application uses the SOTPAR2 to vector quantize acoustic signals from 15 speakers saying the words one through ten. The second application is the prediction of the Mackey-Glass chaotic signal using piece-wise linear approximations.

SOTPAR2 Vector Quantization of Speech Data

The goal of this application is to recognize spoken English digits from one to ten. The SOTPAR2 will be used to vector quantize the sampled frequency representation of each digit. The corpus is a set of 15 speakers saying the digits one through ten. The first 10 speakers are used for training and the last 5 are for testing. The 15 speakers were graduate students and professors in the Electrical Engineering Department at the University of Florida. The speakers represent a wide variety of nationalities and accents, making this task significantly more difficult than one might think. The preprocessing comprised calculating the first 12 cepstral coefficients from 25.6 ms frames overlapped every 12.8 ms (10 kHz sampling). The cepstral coefficients were liftered by a raised sine to control the noninformation-bearing cepstral variabilities for a more reliable discrimination of sounds. These cepstral coefficients were then mean filtered three at a time to reduce the number of input vectors.

The major difference between using the SOTPAR2 VQ method and a standard VQ method is that the SOTPAR2 algorithm is trained to enhance patterns that it was trained with. There are two options to incorporate the temporal characteristics of the SOTPAR2 into this architecture. Typically, one vector quantizer is used to quantize every word in the corpus. This can be done with the SOTPAR2 VQ as well. In this case, the network would need to store all of the temporal information from all ten digits in a single network. Although this is possible, the task of the SOTPAR2 is simplified by training a

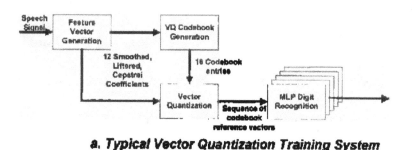

a. Typical Vector Quantization Training System

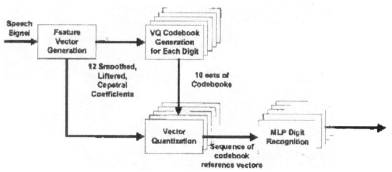

b. SOTPAR2 Vector Quantization Training System

Figure 4-22: Block diagram for the digit recognition system. a) standard digit recognition system, b) SOTPAR2 recognition system

separate SOTPAR2 network for each digit (e.g. each network stores the temporal characteristics of only a single digit). Similarly, we will use a separate MLP to detect each digit. Figure 4-22 shows the overall block diagram of the system.

First we trained the 10 SOTPAR2 VQ networks. This process was done by feeding each network an input consisting of the target digits spoken by the 10 training speakers interspersed with random vectors from the other 9 digits. The training parameters for the network are shown in Table 4-6. The learning rate and neighborhood width are similar to the other networks we have trained. The spatio-temporal parameter β, however, typically takes the shape of a raised sine over the training session. β is different than in the SOTPAR because the neural gas algorithm does not have a spatial structure and the temporal neighborhoods are added externally. The raised sine allows the neural gas PEs to distribute themselves freely at the beginning without temporal interference. In the middle of the training algorithm, after the neural gas PEs are fairly uniformly distributed, the temporal training reaches its maximum and then tails off for fine tuning at the end. The temporal Hebbian increment determines the time constant used to update the temporal weights. If this increment is too large, the network puts too much emphasis on the recent past, and if too small, does not change quickly enough to significantly affect the training. In my experience, an increment of 0.01 seems to work fairly well for most normalized data. Although not required, I also used a conscience term in this network. A conscience algorithm is a standard method of creating a density match between the PEs and the data. If there are a large set of data in one small region of space, an SOM without a conscience may map a single PE to this one region. Including a conscience forces the

SOM to map multiple PEs to this region by forcing each PE to fire at roughly the same rate. A PE that is firing too often is severely penalized such that it cannot fire for a while. The two conscience parameters are typical settings and have little effect on the network.

Table 4-6: Training parameters digit recognition VQ

Training Parameter	Value
Spatio-temporal parameter, β	Raised sine with maximum 0.2
Learning rate, η	Exponential decay from 0.1 to 0.01
SOM neighborhood width	Exponential decay from 5.33 to 0
Number of training epochs	50
Number of PEs	16
Temporal Hebbian increment, $b1$	0.01
Conscience increment, $b2$	0.01
Conscience factor, c	0.2

While training, the activity wavefronts could easily be seen in a plot of the maximum activity in the network over time. This usually picks up the wavefront activity in the network quite well. Figure 4-23 shows the activity of the digit six network with the training data. The instances where the word six is spoken are highlighted between dashed lines. The input data interspersed between the presentations of the 'six'es are random vectors from the other digits. Clearly the activity of the network is much higher when the word six is presented to the network. You should also notice, however, that certain speakers do not adequately match the "global average". For instance, speaker 10, near sample 400, does not create a large activity spike in the network. For larger systems, this can be solved by using multiple networks for each digit, allowing for more variation in the speakers.

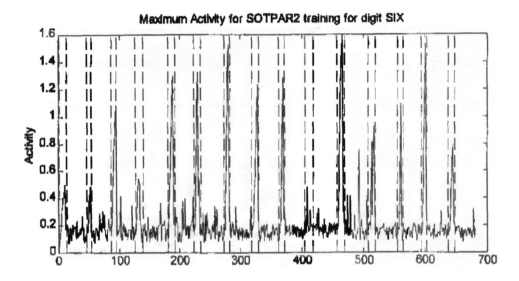

Figure 4-23: Maximum activity in the SOTPAR network for digit 6

After training all 10 models, we verified the validity of the SOTPAR2 networks

by running the digits from each speaker through all 10 models. At each input we plotted

the maximum activity in the network. Figure 4-24 shows the activity in the network for

speaker 14 (one of the test set speakers) with the spoken digit along the horizontal axis

and the SOTPAR2 model (e.g. the vector quantizer for each digit) along the vertical axis.

Notice that the wavefronts are quite noticeable along the diagonal which corresponds to

the proper model responding favorably to the proper digit. Notice also that the

wavefronts for words 6 and 7 both started out high with the pronounciation of the "s"

sound, but this particular speaker did not pronounce the end of either word similar to

those in the training set. Remember, this activity is from an unsupervised, unsegmented,

unlabeled data set created using self-organizing principles. The activity shown in these

networks is helping the proper model remove noise and enhance the signal for later

recognition.

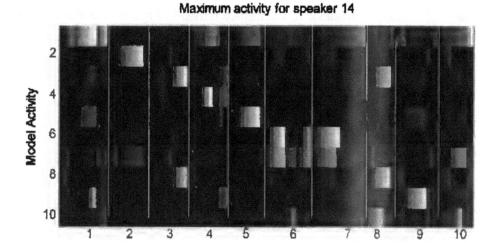

Figure 4-24: Maximum activity in each model for each digit

After the VQ networks were trained, we vector quantized each digit from all 15

speakers, 10 from the training set and 5 from the test set. We also vector quantized the

data using a standard neural gas algorithm so that we could test our results. In order to

remove some of the variability caused by the different rates at which the words and

phonemes were spoken, we passed each sequence of reference vectors from each spoken

word through a gamma memory. The gamma memory has 6 taps and a μ of 0.5 giving it a

depth of 12 samples, which corresponds to the maximum length of any spoken digit in

the corpus (the minimum was 6 vectors). The output of the six taps of the memory is then

fed into an MLP with 6 hidden PEs and a sigmoidal output PE. All 10 networks are

trained simultaneously and a winner-take-all PE is used to select the network with the

largest output. The digit detector with the largest output is declared the winner and this is

compared against the desired signal which indicates which digit was actually spoken.

Table 4-7 shows the training parameters used for each of the MLP subnetworks. These

parameters are based upon well-known rules-of-thumb and variations in these parameters do not greatly affect the performance of the trained network. Figure 4-25 shows a block

diagram of the recognition system.

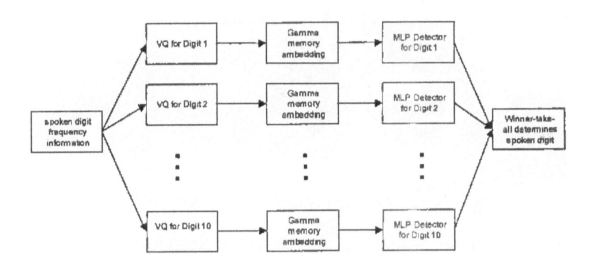

Figure 4-25: Digit recognition system

Table 4-7: Training parameters for each subnetwork of the MLP recognition network

Training Parameter	Value
Hidden layer PEs	6 tanh PEs
Hidden layer learning rate	1
Hidden layer momentum rate	0.7
Output layer PEs	1 logistic PE
Output layer learning rate	0.1
Output layer momentum rate	0.7
Number of training epochs	1000

The entire system was trained with three different sets of data. First, the original

data from the preprocessor was used to train the system. This allows us to validate that

the vector quantization reduces the variability of the data and allows for easier

recognition of the digits. Second, the neural gas algorithm was used to vector quantize the input data. Third, the SOTPAR2 network was used for each vector quantizer. In order to

remove random variations based on initial conditions, each system was trained and tested

5 different times and the results were averaged. Table 4-8 shows the results of the

training. The key figures are the number of misclassifications in the test set. Since MLPs

are universal mappers, a sufficiently large MLP can learn to classify virtually any data

set. A common problem with MLPs is that they can be overtrained, which can be thought

of as memorizing the input instead of finding the features in the input that are important.

If the network is overtrained, it will have very good classification in the training set, but

very poor classification in the test set. Thus, the true indication of performance for MLPs

is the performance in the test set.

Table 4-8: Summary of the digit recognition system performance

System Type	Training MSE	Testing MSE	Training misclass- ifications	Testing misclass- ifications	Percent Correct classification in Test set
No Vector Quantization	0.0005	0.0216	0.0	12.2	75.5
Neural gas VQ	0.0010	0.0321	0.2	10.0	80.0
SOTPAR2 VQ	0.0009	0.0199	0.2	7.6	84.8

The table shows that the SOTPAR2 VQ system reduced the number of errors in

the testing set by 25% over the neural gas VQ system and by 40% in the system without

vector quantization. This performance is due to the reduction in the variability in the

systems. A vector quantization technique removes some of the variability of the signal by

clustering all the inputs and representing every input in the cluster with a single reference

vector. The SOTPAR2 VQ system takes this one step further by using temporal

information to enhance the clustering. The temporal sequence of the feature vectors plays an important role in the vector quantization. For additional comparison, a hidden markov model (IIMM) with 5 states was trained using the original input. The HMM was trained for 50 cycles starting from 5 different initial conditions. The average results for the HMM was 81% correct over the test set.

To further analyze the system, Table 4-9 and Table 4-10 show the "confusion matrices" for two sample systems (one SOTPAR2 VQ system and one neural gas VQ system). The confusion matrix uses a two-dimensional grid to represent the classification of the digits. On the horizontal axis is the classification by the network and on the vertical axis is the correct classification. Thus, perfect classification contains all the values on the diagonal of the matrix. Misclassifications show up as non-diagonal terms. The non-diagonal terms show which digits are being confused. For example, the first row in Table 4-9 shows the networks classifications for the 5 instances of the digit one in the test set. Four of these "ones" were correctly classified and the fifth was classified as a five. These confusion matrices show interesting details about the classification of the network. For instance, nine is confused with one and five because one has the same 'n' sound as nine and five has the same 'i' sound as nine. Similarly, the 'n' sound in seven and ten also tends to confuse the networks and they misclassify them as one as well. The neural gas VQ systems performed very poorly when presented the digits nine and ten.

In summary, the SOTPAR2 VQ system performed significantly better than the static VQ system and the system without vector quantization. The temporal plasticity of the algorithm allowed for better quantization performance and noise resistance.

Table 4-9: Confusion matrix for SOTPAR2 VQ system

Input Digit		Network Classification									
		1	2	3	4	5	6	7	8	9	10
	1	4				1					
	2		5								
	3			5							
	4				5						
	5					5					
	6						5				
	7	1				1		3			
	8								5		
	9	1				1				3	
	10	1									4

Table 4-10: Confusion matrix for neural gas system

Input Digit		Network Classification									
		1	2	3	4	5	6	7	8	9	10
	1	4				1					
	2		3					2			
	3			5							
	4				5						
	5					5					
	6						5				
	7							5			
	8								5		
	9	1				1		2		1	
	10	1					1	1	1		2

Time Series Prediction

In this application we compare the performance of the neural gas algorithm to the

SOTPAR2 algorithm for prediction. The application is based on the study of Martinetz,

et.al. [Mar93] where they used the neural gas algorithm to do time series prediction. They

partition the input space into regions with each region having its own local linear predictor. Thus, each Voronoi region of the input space has a single linear predictor.

They used the Mackey-Glass time series:

$$\dot{x}(t) = \beta x(t) + \frac{\alpha x(t-\tau)}{1-x(t-\tau)^{10}}$$

with the parameters $\alpha = 0.2$, $\beta = -0.1$, and $\tau = 17$. X(t) is quasi-periodic and chaotic signal

with a fractal attractor dimension 2.1 for the parameters chosen here. The characteristic

time constant of x(t) is 50 which makes it particularly difficult to forecast x(t+δt) with

δt>50 [Mar93]. The time series is predicted using a 4 dimensional embedding with a time

lag of 6 between each input, thus the predictor is trying to estimate:

$$x(t+6) \approx f(v) \text{ where } v = x(t), x(t-6), x(t-12), x(t-18)$$

For each neural gas node i there is one linear predictor with coefficients y_i and a_i

defined by:

$$\tilde{y} = y_i + a_i \cdot (v - w_i)$$

where $\tilde{y}$ is the prediction of the next point in the sequence. The equation contains an

average prediction for this region, y_i, and an offset from this average prediction based on

the distance between the reference vector and the input vector. The vector a_i has 4

components, one for each input dimension. Using an LMS approximation to gradient

descent, the update equations for the y's and a's are as follows:

$$err = y - y_i - a_i \cdot (v - w_i)$$
$$\Delta y_i = \eta h_\lambda(k_i(v,w)) \cdot err$$
$$\Delta a_i = \eta h_\lambda(k_i(v,w)) \cdot err \cdot (v - w_i)$$

where err is the prediction error, η is the learning rate, h_λ is the neighborhood function, and k_i is the neighborhood ranking (the updates are done in a manner similar to the neural gas algorithm itself).

<u>Results</u>

For most signals, predicting the next value in a sequence given the previous values of the sequence is actually fairly simple. Most signals are continuous and vary fairly slowly, so the prediction can be done very accurately with very small models. The key to dynamical prediction, however, is how well the model has captured the dynamics of the system. One of the best ways to test this is to implement multi-step prediction. Multi-step prediction is implemented by initializing the memory of the system to a valid point in the trajectory of the system dynamics and then predicting the next point in the system. This predicted point is then fed back to the input of the system and is used to predict the next point in the system. Small errors in the predictions that are fed back and used to make even more predictions can often produce very large errors after a few iterations. If the network has actually captured the system dynamics, then it should be able to start at any point in the trajectory of the system and follow that trajectory autonomously for a certain number of steps. The distance that the network can follow a trajectory is dependent on the system that is being modeled. For instance, a chaotic system can be modeled only for a short period of time before the smallest of errors drives the network off of the correct trajectory. In fact, this is one possible definition for chaotic systems – systems where small changes in initial conditions create large changes in the system dynamics.

The static neural gas algorithm and the SOTPAR 2 algorithm were both trained for 100 epochs over a 500-point segment of the Mackey-Glass time series. The time series was generated using a 4th order Runge-Kutta algorithm. We first trained the PE locations and then trained the linear predictors, although these can easily be done simultaneously. Table 4-11 shows the training parameters for both the SOTPAR2 and neural gas VQ networks. The only difference between the two networks is that the neural gas system does not use the spatio-temporal parameter or the temporal Hebbian increment parameter. Over the training period, the SOTPAR2 algorithm consistently produced better prediction results than the neural gas. The final average one-step prediction error for the two networks was 7.63×10^{-6} for the SOTPAR2 algorithm and 9.15×10^{-6} for the neural gas algorithm. The SOTPAR2 reached a MSE which was 16% lower than the neural gas, but as we said previously, the single-step prediction error does not determine how well the dynamics of the system have been captured. The final maximum single step prediction error may provide a better indication of how well the system will perform under multi-step prediction. The SOTPAR2 maximum single step prediction error was 1.36×10^{-4} while the neural gas had a maximum error of 5.86×10^{-4}. The SOTPAR2's maximum error was 76% lower than the neural gas's maximum error. Since large errors will quickly drive the network away from the desired trajectory of the chaotic system, the maximum error is more relevant than the MSE.

Next I tested the networks using multi-step prediction. After each training epoch, the weights were frozen and 50 randomly selected initial conditions were loaded into the networks. The predicted outputs were fed back into the networks to predict more values.

Table 4-11: Training parameters for SOTPAR2 and Neural gas VQ for prediction

Training Parameter	Value
Spatio-temporal parameter, β	Raised sin with maximum 0.2
Learning rate, η	Exponential decay from 0.1 to 0.01
SOM neighborhood width	Exponential decay from 16.7 to 0
Number of training epochs	50
Number of PEs	50
Temporal Hebbian increment, $b1$	0.01
Conscience increment, $b2$	0.01
Conscience factor, c	0.2

The MSE between the desired trajectory and the 50 autonomous trajectories generated by the networks was then computed. The top plot in Figure 4-26 shows the MSE versus training epoch for both the neural gas (dotted line) and SOTPAR2 (solid) for 10 step prediction on a semilog scale. The bottom plot shows the average performance

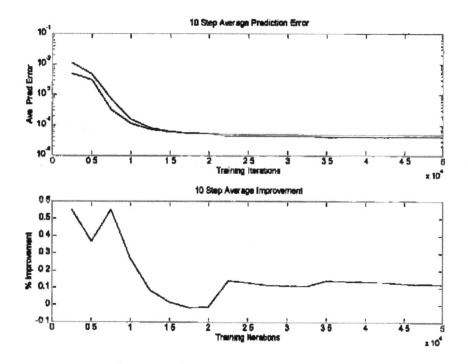

Figure 4-26: 10-step prediction error and average improvements for the neural gas (dashed line) and SOTPAR2 (solid line) predictors

improvement between the neural gas and SOTPAR2 versus the training time

$(\frac{E_{NGas} - E_{SOTPAR2}}{E_{NGas}})$. Initially, the SOTPAR2 trained faster than the neural gas, but in the

end the SOTPAR2 only performed 10% better on average than the neural gas.

As the prediction length increases, the ability of the SOTPAR2 to use temporal

information to remove noise in the signal should improve the performance of the system.

The embedding of the data creates a state space representation of the system output. This

can be visualized as a trajectory moving through an N-dimensional space. The networks

are carving out regions of this N-dimensional space and one predictor is assigned to each

Voronoi region. When the current position of the trajectory of the system (either the

original system or the autonomously running network trying to predict the system) ends

up in a certain region, that region's predictor is used to predict the next output. The neural

gas algorithm determines which region the trajectory is in using only the static

information available to it. The SOTPAR2 algorithm uses the past locations of the

trajectory as well to help guide the selection of the proper Voronoi region. Thus, if noise

in the system has moved the trajectory from the correct region to a neighboring region,

the SOTPAR2 algorithm uses its stored temporal information (obtained from the training

data) to select the more appropriate predictor. Remember, that even if the original signal

is "noise-free", the prediction of the signal will induce a significant amount of noise into

the reconstruction of the trajectory. Figure 4-27 shows the results for the 25-step

prediction. Notice now that the SOTPAR2 is beginning to significantly outperform the

neural gas, providing a 60% decrease in the prediction error.

144

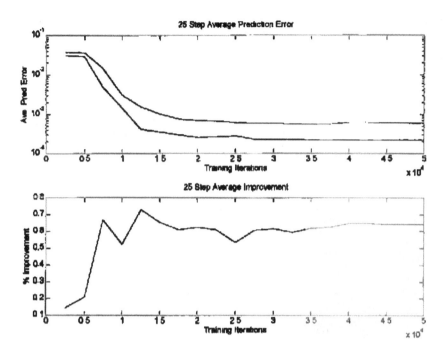

Figure 4-27: 25-step prediction error and average improvements for the neural gas
(dashed line) and the SOTPAR2 (solid line) predictors

Figure 4-28 and Figure 4-29 show the performance of the two systems for 50

steps and 100 steps. The 50-step prediction still shows a significant decrease in the MSE

for the SOTPAR2 algorithm as it performs 45% better. The 100-step prediction,

however, shows that we have reached the maximum capabilities of the network. This may

not be the result of poor network performance, however. Chaotic systems can only be

predicted accurately in the short-term. Even with identical systems, a very small change

in the initial conditions produces large long-term errors. As mentioned previously,

predicting beyond 50 steps is very difficult for this data set. To further illustrate this

point, I calculated the Lyapunov exponent for our segment of the Mackey-Glass signal.

The Lyapunov exponent is a dynamical system parameter that measures the average rate

of divergence for nearby trajectories. They can be thought of as the eigenvalues of the local linearization of the dynamics.[Ger94] If any Lyapunov exponent is greater than zero, then it is a chaotic signal. In our case, the largest Lyapunov exponent was approximately 0.067 bits per second. With any small difference between two sets of initial conditions, two trajectories of a chaotic system will diverge at a rate approximately equal to

$$d_n = d_0 2^{n\lambda}$$

where d_n is the distance between the two trajectories at time n, d_0 is the initial distance between the two trajectories, n is the number of iterations between the initial and final positions, and λ is the dominant Lyapunov exponent. Using this equation and the average MSE for the SOTPAR2 predictor, Figure 4-30 shows the trajectory divergence for the input data. The horizontal line in the figure indicates the variance of the input signal and

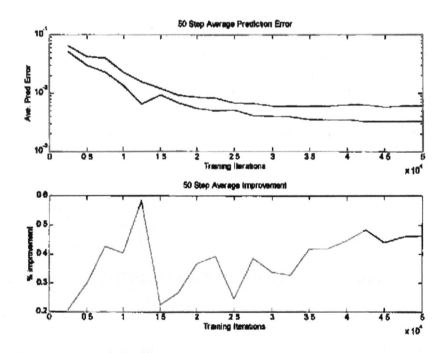

Figure 4-28: 50-step prediction error and average improvements for the neural gas (dashed line) and SOTPAR2 (solid line) predictors

signifies the practical limit of predicting the signal with these parameters. Any predictor with an MSE greater than the variance is no better than simply predicting the mean of the signal at all times. Thus, for all practical purpose, this signal cannot be predicted accurately beyond approximately 60 samples with the given mean squared error. Additional training could reduce the MSE further, but since the slope of the divergence is so great near 60 samples, it will not improve the predictability of the signal significantly.

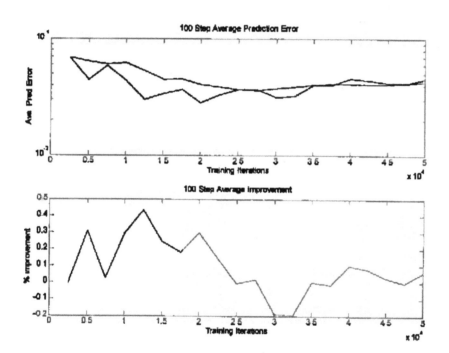

Figure 4-29: 100-step prediction error and average improvements for the neural gas (dashed line) and SOTPAR2 (solid line) predictors

Lastly, I tested both systems by determining their average prediction length (number of predictions before the error was greater than 0.001). On average, the SOTPAR2 could predict 30% further on average than the neural gas algorithm. Table 4-12 shows the results that were collected over 8 different training runs with 40 different starting points for each training run. The table also shows the dependence that the prediction length has on β, the spatio-temporal parameter. As β increases beyond 0.10 the

performance decreases due to the overemphasis on temporal information.

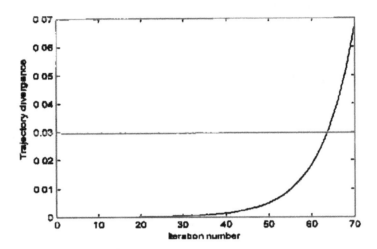

Figure 4-30: Trajectory divergence for Mackey-Glass prediction

Table 4-12: Average prediction length for the neural gas and SOTPAR2 predictors with different values of β for the SOTPAR2

Model	Ave. Prediction Length
Neural gas	33
SOTPAR2 β=0.05	40
SOTPAR2 β=0.10	43
SOTPAR2 β=0.15	39
SOTPAR2 β=0.20	33

Summary of chaotic prediction

In this application we have shown how the temporal plasticity of the SOTPAR2

can help disambiguate the static spatial information of the multi-step predictor. The

dynamic vector quantization helps reduce the variability inherent in the input by

anticipating (based on training) the future inputs. Specifically, the improvements can be attributed to two factors. First, the SOTPAR2 trains faster because the vector quantization is based on a trajectory in state space, not just the current location in state space. Secondly, the SOTPAR2 improves the multistep prediction because it can help remove the noise created by the iterative predictions. Both of these factors are due to the the temporal plasticity in the network that causes the Voronoi regions to shrink and grow based upon the previous inputs and training. The anticipation inherent in the dynamics of the network allow for a better selection of the linear predictor.

Dynamic Subgrouping of RTRL in Recurrent Neural Networks

This section presents the applications that were used to test the dynamic subgrouping methodology to train recurrent neural networks. The activity diffusion and temporal Hebbian learning concepts allow for the temporal self-organization of the PEs in a fully recurrent network. Using the temporal self-organization, many of the gradient calculations in the RTRL algorithm can be ignored, reducing the computational requirements from $O(N^4)$ to $O(N^2)$. The first application is the system identification of a nonlinear system with memory. The second application is also system identification applied to the nonlinear passage dynamics in a noise cancellation problem.

System Identification

The first experiment will evaluate the identification of the nonlinear system with memory presented in [Cam95]. System identification is a common task in control theory where the input-output characteristics of an unknown system are modeled as accurately as possible. This is accomplished by injecting the same signal (typically random noise if possible) into both the unknown plant and the model. The model is then trained to mimic the response of the plant by using the difference between the plant output and the model

output as the criterion to minimize. Figure 4-31 shows a block diagram for system

identification. For simplicity, often the random input is first injected into the plant and the

plant's input and output are stored so that the model can be trained off-line. ANNs

provide a powerful methodology for system identification because they are universal

mappers and thus in theory can model any plant.

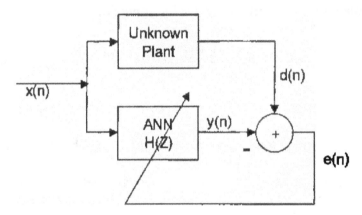

Figure 4-31: System identification block diagram

The system we will model is described by the following equations:

$$z(t) = 0.0154x(t) + 0.0462x(t-1) + 0.0462x(t-2) + 0.0154x(t-3)$$
$$+ 1.99z(t-1) - 1.572z(t-2) + 0.4583z(t-3)$$
$$y(t) = \sin[z(t)]$$

This system uses the past three values of the state z and the present and past three values

of the input to update the value of the state z. This state output z is then passed through a

sine function to determine the system output. A sequence of 1000 points of random

uniform noise between −1 and +1 was generated and passed through the system

equations. This set of input-output data was then used to train the network. Each 1000

points is an epoch of training. This data was used to train 4 networks using different

150

learning mechanisms: the full RTRL algorithm, the Zipser static subgrouping algorithm, the dynamic subgrouping algorithm using the full π matrix to determine the subgroups, and the dynamic subgrouping algorithm using the first-order estimate of the π matrix to select the subgroups. The results of each network were averaged over 5 different training runs with different sets of initial conditions (the same 5 sets of initial conditions for each network).

Figure 4-32 shows the learning curves for the 4 different networks when the subgrouping algorithms used two groups of 3 PEs (six total PEs). As expected, the full RTRL and the dynamic subgrouping with π (DS-π) algorithms were very similar. Remember that the dynamic subgrouping with π algorithm produces the best possible results for each particular configuration. This indicates that the chosen network setup (6

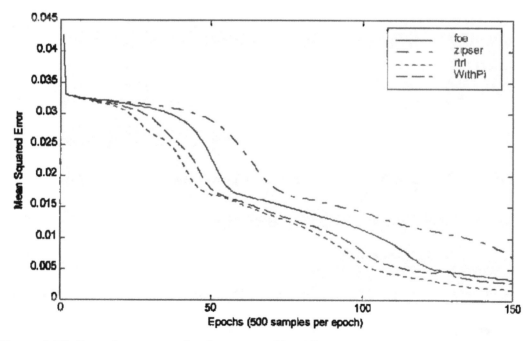

Figure 4-32: Learning curves for the system identification problem using the RTRL, Zipser, DS-π, and DS-FOE algorithms

151

PEs with 2 groups) is sufficient to train the network properly. The dynamic subgrouping using the first-order estimate (DS-FOE) produced slightly worse performance than the full RTRL and DS-π algorithms, but required significantly less time per epoch to train. Zipser's algorithm took the most epochs to train. The DS-FOE algorithms and Zipser's algorithm both require $O(N^2)$ operations while the full RTRL and DS-π algorithms require $O(N^4)$ operations.

Table 4-13 shows the time per epoch, the number of epochs to reach an error of 0.01, and the overall time required to reach an MSE of 0.01. The timing was done on a Pentium 166MMX PC using MATLAB code and the Mathworks Visual Matlab compiler. The compiling of the code helps reduce the senstitivity of the algorithms to the actual code, but there is certainly some variation between these results and code that is fully optimized and/or written in C. In theory, the DS-FOE algorithm will be N^2/m^2 faster and Zipser's algorithms will be g^2 faster than the full RTRL algorithm where N is the number of PEs, m is the number of temporal neighbors, and g is the number of subgroups. When the number of temporal neighbors divides evenly into the number of PEs, Zipser's algorithm and the DS-FOE algorithm will have the exact same performance. Since we have 2 subgroups in this case (g=2, m=3), both algorithms should be 4 times faster and they were approximately 3 times faster in reality (more on this later). Since the full RTRL trained in fewer epochs the actual performance improvement was only 44% for Zipser's algorithm and 57% for DS-FOE. For larger networks, however, the training time per epoch improvement will dominate the total training time and the DS-FOE algorithm will drastically decrease the overall time required to train the network. For instance, with a network as small as 24 PEs using 3 temporal neighbors, the DS-FOE

algorithm will execute one epoch of training 32 times faster than the full RTRL.

Table 4-13: Training times for the four algorithms for the system identification problem

	Time per Epoch	Num Epochs to 0.01	Total Training Time (secs)	% Improvement
Full RTRL	96	88	8,448	0%
Zipser	33	140	4,620	44%
DS-FOE	33	110	3,630	57%

Comparison of the Number of Neighbors

Figure 4-33 shows the learning curves for the system identification problem using

different numbers of neighbors. Every network had a total of 6 fully recurrent PEs, but

used between 2 and 6 temporal neighbors. If 1 neighbor is used, the algorithm defaults to

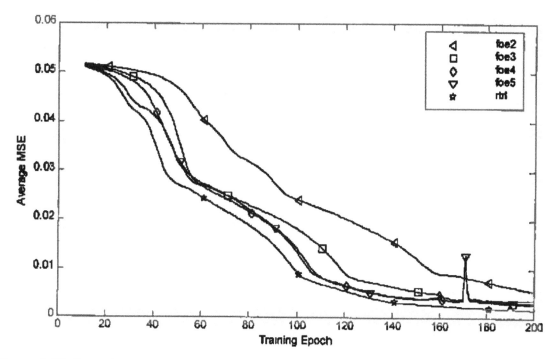

Figure 4-33: Learning curves for the DS-FOE algorithm using 2,3,4,5, and 6 neighbors

153

a first-order estimate of the RTRL algorithm where the only long-term gradient information comes from within the self-recurrent loops. When 6 neighbors are used, the algorithm is identical to the full RTRL algorithm. As expected, the learning curve with only 2 temporal neighbors was the slowest. The more temporal neighbors are added, the faster the convergence. In this particular application, the law of diminishing returns seemed to take effect after 3 or 4 temporal neighbors.

Next, I studied the training times per epoch to determine if the theoretical speed-up estimates are correct. Table 4-14 summarizes the results. The left column shows the number of temporal neighbors used in each network (using a 6 PE network). As mentioned before, the dynamic subgrouping algorithm with one temporal neighbor is nearly a static approximation to the RTRL algorithm. The only temporal gradient information comes from the recurrent loop in the network. Thus, this network will require the least time per epoch but will generally require many more epochs (if it trains at all) to be properly trained. Using 6 temporal neighbors is identical to using the full RTRL algorithm. The table shows that the N^2/m^2 speed-up estimate is not very accurate, especially when the number of temporal neighbors is low. A quick analysis of the algorithm showed that the actual speed-up is closer to $N*(N+3)/m*(m+3)$. The number 3 comes from the fact that each weight update requires 3 extra computations. When the number of temporal neighbors is close to 3, the (m+3) term in the denominator has a large impact on the speed-up. The 4th column of the table shows that this new estimate is much closer to the actual timed values. Figure 4-34 shows that as the number of temporal neighbors increase, the estimates and actual values become much closer.

Table 4-14: Training times and theoretical speed-up for DS-FOE algorithm

Number of Neighbors	Actual Time per Epoch	$O(N^2/m^3)$ speed-up estimate	Precise speed-up estimate	Num Epochs to MSE<0.01
1	10	3	8	475
2	21	11	18	155
3	33	24	32	110
4	50	43	50	100
5	70	67	71	98
6/RTRL	96	96	96	88

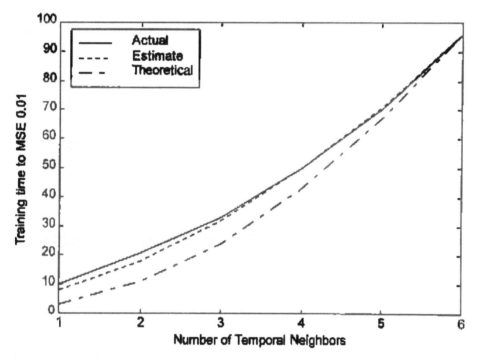

Figure 4-34: Execution time per training epoch for a 6 PE RMLP using DS-FOE and varying the number of temporal neighbors

Finally, Figure 4-35 shows the overall execution time to train each network to a mean squared error of 0.01 when the number of temporal neighbors was varied. The plot clearly shows that using either 2 or 3 temporal neighbors minimizes the overall training time of the network. One neighbor did not contain enough temporal gradient information

157

Five hundred points of random noise were injected into the system and the input and output were collected. A 6 PE RMLP was trained with the full RTRL algorithm, Zipser's algorithm, the DS-π algorithm, and the DS-FOE algorithm. Figure 4-37 shows the learning curve and the semi-log plot of the learning curve for the four algorithms. Each learning curve is averaged over the same four sets of random initial conditions. As expected, the DS-π algorithm again performed virtually the same as the full RTRL algorithm – this proves that the problem can be solved using only the full gradient information from at most 3 temporal neighbors of each PE. Zipser's algorithm again performed the worst of all algorithms and the DS-FOE algorithm performed between

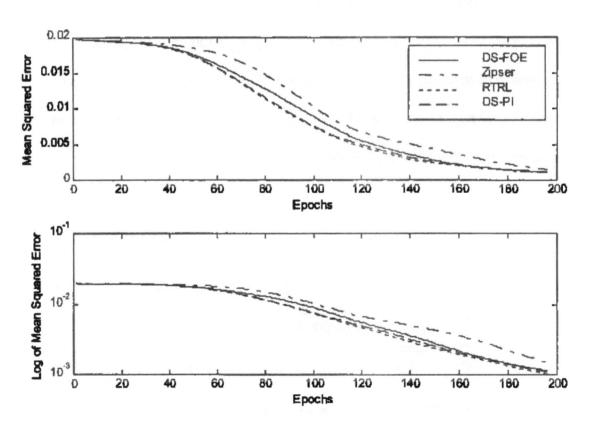

Figure 4-37: Learning curves and the log of the learning curves for the passage dynamics system identification

Zipser's and the full RTRL algorithm. The DS-FOE reached the bottom of the learning

curve significantly faster than Zipser's method. Also, although the learning curves all

look very similar, since they all have a shallow slope, it will take Zipser's algorithm

significantly longer to reach the same MSE as the DS-FOE algorithm.

Figure 4-38 plots the elements of the 6th row of the Z matrix over time for the

passage dynamics problem. This plot shows the amount of temporal information shared

between PE6 and all the other PEs. By choosing the 2 PEs (along with the self-recurrent

loop between PE6 and itself) that have the highest temporal correlation with PE6, the

algorithm trains in nearly the same number of epochs with much less computation per

epoch. This produces a much lower overall training time. This particular application does

not have very interesting switching dynamics. At first, the winning neighbors were PE3

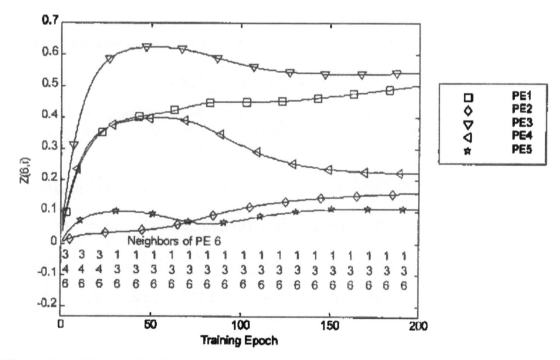

Figure 4-38: Temporal neighbors of PE 6 for the passage dynamics problem

and PE4 but after 30 epochs, PE1 replaced PE4. Again, as expected, this plot verifies that the temporal gradient information content between PEs is smooth and slowly changing. This allows us to subgroup the PEs without switching too often which may disrupt the training.

For comparison, the same input was tested using 4 other common networks. First, a linear combiner (FIR) with 50 taps was tested. Since the passage dynamics are nonlinear, it came as no surprise that the linear combiner could not model the system and the best it could do was achieve an MSE of 0.10. Next, a tap-delay neural network (TDNN) was tested. The TDNN is a nonlinear version of the FIR where a tap-delay line is added at the beginning of an MLP, thus giving the MLP the ability to map temporal signals. The TDNN was setup with 7 taps and 6 hidden PEs so that both it and the RMLP would have the same number of weights (48). The TDNN trained more quickly than the RMLP architecture. The average training time was 40 epochs versus approximately 100 epochs for the RMLP. The average MSE after 200 training epochs, however, was twice that of the RMLP. The RMLP is a more powerful architecture than the TDNN and thus was able to capture more of the dynamics of the system. The TDNN simply does an embedding of the input into a multi-dimensional static pattern which then must be mapped by the MLP. The limited flexibility of the memory in the TDNN prevents it from being applied to some complex problems. Using a gamma memory in place of the tap-delay line did not significantly improve the performance of the TDNN.

Lastly, I trained an identical RMLP with BPTT. BPTT is a trajectory algorithm for computing the temporal gradients. This means that the network is updated in a batch mode after each trajectory has been processed in both the forward and backward direction. Since batch mode algorithms train more slowly than on-line algorithms at the same step size (the gradients are already being averaged), the step size for the BPTT algorithm was set larger than the RMLP step size. The BPTT algorithm still required over 2000 epochs to properly train the network. The trajectory size was varied between 5 and 50 samples (the depth at which the gradients are calculated) without significant improvement in the performance of the BPTT algorithm.

Summary of Dynamic Subgrouping

This section presented two control applications that were solved with a recursive multi-layer perceptron and the forward propagation of gradients. The DS-π algorithm showed that with the proper selection of neighbors, a dynamic subgrouping algorithm can greatly decrease the required computation per epoch while consistently training in nearly the same number of epochs as the full RTRL algorithm. The DS-FOE algorithm showed that with a simple first order approximation to the π matrix, the neighbors could be selected well enough to consistently outperform Zipser's algorithm.

This is the last demonstration of the temporal organization of PEs using diffusion and temporal Hebbian learning concepts, and it may be the most important because it is a dramatic improvement on a generically applied method. Any methodology that uses real-time recurrent learning to train a neural network can use this methodology and expect significant performance improvements with little loss of power. The methodology reduces the number of computations by a factor of N^2/m^2, where N is the number of PEs in the network and m is the number of temporal neighbors used. When the size of the groups is held constant as the networks grow in size (larger networks mean more subnetworks), which practically and theoretically makes sense, the overall computational

requirements for the DS-FOE algorithm are only $O(N^2)$ versus $O(N^4)$ for RTRL.

CHAPTER 5
CONCLUSIONS AND FUTURE RESEARCH POTENTIAL

Conclusions

The goal of this work was to provide a new mechanism that allows neural networks to process temporal patterns and signals more efficiently. Inspired by biology, the fundamental concept underlying this proposed mechanism is the temporal self-organization of PEs in a network. The temporal organization of the PEs allows them to process the temporal data in a more organized and structured manner. The fact that it is done in an unsupervised (or self-organizing) manner means that the temporal organization can be considered independent of the normal operation of the network. The fundamental operation of the underlying neural network architecture is unchanged, allowing for the simple integration of the method with the existing algorithms. This is my *main theoretical contribution – that temporal self-organization in space-time can be easily added to existing neural network architectures with little change in their fundamental operation, yet significantly improve their performance on temporal applications.*

Most temporal neural networks simply add short-term memory to an existing static neural network. My methodology can be viewed as a unique combination of short-term and long-term memory. The temporal information from the input is stored in both the activations of the network and also the structure of the network. This creates a *unique trainable memory architecture that responds favorably to signals similar to those that it was trained with.* This memory is more similar to biological memories. *Activity diffusion is the local mechanism used to transfer information throughout network.* It is biologically inspired and does not require global connections or communication. Activity diffusion and temporal Hebbian learning are used to train the temporal components of the network. These techniques were applied to three different neural network architectures and increased the performance of each of them. There is no reason to believe that this mechanism could not also be applied to many other architectures.

When applied to a self-organizing map, the activity diffusion concept creates a spatio-temporal mapping in the network. The SOM maps similar input vectors to similar locations in the space of the SOM, and the SOTPAR dynamics map temporally similar inputs (e.g. successive inputs) to similar locations in the SOM. The combination of these two concepts creates the unique concept of self-organization in space and time. When the architecture processes its data, the activity diffusion creates waves of temporal activity that travel through the output map and influence the future operation of the network. These traveling waves provide a truly unique memory mechanism to the system. This architecture can be trained in an unsupervised manner with unsegmented data and will map common sequences in the input to sequential PE locations in the SOM.

When applied to a landmark recognition problem in robotics, the SOTPAR architecture was able to use the temporal information in the signals to smooth the noisy turning data and gracefully handle the time-warping caused by variations in the speed of the robot. A specially modified version of the SOTPAR provided one-shot training, landmark recognition, endpoint detection, and location information. This neural implementation of a landmark recognition system is unique in the robotics field. Real data from a robot was collected and tested by Pedro Kulzer. The results were very promising but testing was discontinued after he graduated. More research, testing and comparisons to other methods are required but may provide a new and powerful mechanism for robotic navigation.

The SOTPAR was also applied to the mapping of phoneme sequences. Phoneme sequences from an industry standard word database were mapped to an output space using traveling wavefronts with a pre-specified width. This mechanism creates local clusters of PEs that represent variations of a single phoneme, but the local clusters are globally organized by their temporal sequence. When compared against standard vector quantization techniques, the global temporal ordering created a much more organized map with most instances of each phoneme being clustered in a single small region of the output map. This architecture has been implemented in hardware and interactively demonstrated at neural network conferences by the German research group headed by Ruwisch.

Next I applied the self-organization in space-time method to a neural gas architecture which freed the network from the spatial lattices of the SOM. The SOTPAR2 architecture uses a secondary connectivity matrix to store the temporal information from the training data. The connectivity matrix is trained using temporal Hebbian learning and the dynamics of the activity diffusion are very similar to the SOTPAR. As each PE fires, the activity from this PE diffuses over time and space to affect the future operation of the network. In this case, however, the activity diffuses through the connectivity matrix which allows us to train the temporal neighbors more effectively. The concept that best describes the training and operation of this structure is *anticipation*. The network is again trained in an unsupervised manner, without segmentation or labeling of the data. The network is trained so that it anticipates future inputs based on the past inputs (and its training) and responds favorably to these anticipated inputs. When applied to a vector quantization algorithm, the *network uses temporal information to remove spatial noise in the input by creating dynamic Voronoi regions that shrink or grow based upon the past of the signal*. This concept is unique in the neural network community. For instance, if the previous 3 inputs to a network were the phonemes 's', 'm', and 'a', then the phoneme 'l' would be much more likely to be accepted next than the phoneme 'd'.

The SOTPAR2 architecture was applied to the vector quantization of speech data for a digit recognition problem. The input data consisted of 15 graduate students and professors speaking the words one through ten. The first 10 speakers were used as a training set and the last 5 were used as a test set. The SOTPAR2 and neural gas algorithms were trained and used to vector quantize the speech data before being processed by a set of MLP digit detectors. The SOTPAR2 VQ methodology reduced the number of errors in the test set by 25%. The SOTPAR2/MLP architecture also outperformed a HMM system by 20%. These results are promising, but a large scale effort is required to verify the usefulness of these techniques.

The SOTPAR2 was also used to predict the Mackey-Glass chaotic signal. The SOTPAR2 was used to cluster the input trajectories into regions, each of which used a local linear predictor. This prediction system was compared against one that used the static neural gas algorithm. The SOTPAR2 predictor reduced the mean squared error of a 25-step iterative prediction by over 60%. On average it was able to predict 30% farther before reaching an error greater than 0.01. The temporal information in the SOTPAR2 created a dynamically changing vector quantization algorithm that allowed for better selection of the local linear predictors based upon the past information in the signal.

Lastly, the activity diffusion and temporal correlation concepts were applied to the temporal training of recurrent networks. Recurrent networks have been largely ignored in the past because of the difficulty in training them. As the other methodologies are pushed beyond their capabilities, the recurrent networks are starting to be used more and more. Using the temporal self-organization concepts inside the real-time recurrent learning algorithm allows the gradient information to be clustered or subgrouped. This *subgrouping of the gradients reduces the number of cross-terms that must be computed in the RTRL algorithm and greatly reduces the number of operations required to train the network*. In a fully recurrent network, the number of operations is reduced from $O(N^4)$ to $O(N^2)$. The dynamic subgrouping greatly enhances the performance of this subgrouping technique over the static arrangement proposed by Zipser. Because the use of recurrent neural networks has been dominated by control applications, we applied this architecture to two control applications, the system identification of a nonlinear dynamic system and to the passage dynamics of a noise cancellation system. For both systems, the computational speed of each training epoch was significantly faster than the full RTRL algorithm, and the number of epochs required for training was not signficantly greater. On a small network with only 6 PEs, my dynamic subgrouping method could train the network over twice as fast as the RTRL method. On larger networks, the $O(N^4)$ term will make the results even more impressive.

The methodology of using activity diffusion and temporal Hebbian learning to self-organize neural networks in space and time seems to have great potential. Each network I applied the network to obtained properties slightly different than the others and unique in the field of neural networks.

Future Directions

There are many possible directions for future research in this area – in fact, it was difficult to stay focused on only these three architectures as more and more possible applications came to mind. The SOTPAR and SOTPAR2 algorithms are a unique methodology for mapping spatio-temporal patterns and better methodologies need to be determined to use these architectures for purposes other than vector quantization. The dynamic subgrouping algorithm uses a first-order approximation to the matrix of gradients to determine which PEs should be temporal neighbors. This method works well, but is suboptimal and could work better. New grouping criteria can be determined that will improve the performance even more. Additionally, this methodology should be studied in regards to the Extended Kalman Filter methodology of using second-order information to train neural networks. Most EKF implementations use similar concepts to Zipser's in that they ignore cross terms for the second-order information. This static disposal of gradient information should easily be replaced by a dynamic subgrouping similar to ours. Additionally, our method can be even more easily applied to this algorithm since the EKF algorithm already requires the first-order gradient information which can be obtained using the dynamics subgrouping technique.

In a broader sense, the temporal self-organization technique could be applied to other architectures. For instance, it may be able to be used in the feedforward portion of a recurrent neural network to prune unneeded connections. It could also be used in MLPs or TDNNs to organize the hidden layer PEs. In general, one of the main difficulties with the MLP or TDNN is that the PEs are all independent. Without any outside organization, each PE attacks the problem as if it were the only PE (e.g. trying to solve the largest source of error). If temporal ordering were imposed on these PEs, the network may be able to achieve a better global learning style, like divide-and-conquer. The concept of self-organization in space and time may provide many important improvements in the ability of neural networks to solve temporal problems.

REFERENCES

[Ans94]   B. Ans, Y. Coiton, J-C. Gilhodes, J-L. Velay, A neural Network Model for
          Temporal Sequence Learning and Motor Programming, *Neural Networks*,
          Vol. 7, no 9, pp 1461-1476, 1994.

[Bou90]   H.A. Boulard, How Connectionist Models Could Improve Markov Models
          for Speech Recognition, in R. Echmiller, ed. *Advanced Neural Computers*,
          pp. 247-254, Amsterdam: North-Holland, 1990.

[Bur93]   N. Burgess, J. O'Keefe, and M. Recce, Using Hippocampal "Place Cells
          for Navigation, Exploiting Phase Coding", in *Advances in NIPS 5*, Morgan
          Kauffman, San Mateo, 1993.

[Bur95]   N. Burgess, M. Recce, and J. O'Keefe, Hippocampus – Spatial Models, in
          M.A. Arbib (Ed), *The handbook of Brain Theory and Neural Networks*,
          Bradford books/MIT Press, 1995.

[Cam95]   P. Campolucci, F. Piazza, A. Uncini, Online Algorithms for Neural
          Networks with IIR Synapses, *IEEE Int. Conf. on Neural Networks ICNN-
          95*, Perth, Australia, Dec. 1995.

[Cha93]   G.J. Chappell and J.G. Taylor, The Temporal Kohonen Map, *Neural
          Networks*, Vol. 6, pp. 441-445, 1993.

[Chr93]   C. Christodoulou, G. Bugmann, T.G. Clarkson, J.G. Taylor, The Temporal
          Noisy-Leaky Integrator Neuron Model, in *Recent Advances in Neural
          Networks*, R. Beale (ed), Ellis Norwood Publishing, 1993.

[Chr95a] C. Christodoulou and T. Clarkson, A Review on the Stochastic Firing
          Behaviour of Real Neurons and How It Can Be Modelled, in J. Mira and
          F. Sandoval (eds), *From Natural to Artificial Neural Computation*,
          Lecture Notes In Computer Science, Springer-Verlag, 930, 223-230, 1995

[Chr95b] C. Christodoulou, T. Clarkson, and J.G. Taylor, Temporal Pattern
          Detection and Recognition Using the Temporal Noisy Leaky Integrator
          Neuron Model with the Postsynaptic Delays Trained Using Hebbian
          Learning, in *Proceedings of the World Congress on Neural Networks*,
          1995.

[Cri94]   D. A. Critchley, Extending the Kohonen Self-Organising Map by Use of
            Adaptive Parameters and Temporal Neurons, Ph.D. Thesis, University
            College London, Department of Computer Science, February 1994.

[Cun94]   R.K. Cunningham and A.M. Waxman, Diffusion-Enhancement Bilayer:
            Realizing Long-Range Apparent Motion and Spatiotempoarl Grouping in
            a Neural Architecture, *Neural Networks*, Vol. 7, Nos. 6/7, pp. 895-924,
            1994.

[Cyb89]   G. Cybenko, Approximation by Superpositions of a Sigmoidal Function,
            Mathematics of Control, Signals, and Systems, Vol. 2, pp. 337-341, 1989.

[DeV91]   B. de Vries and J.C. Principe, A Theory for Neural Networks with Time
            Delays, in R.P. Lippman, J. Moody, & D.S. Touretzky (Eds.), *Advances in
            Neural Information Processing Systems 3 (pp. 162-168)*, Morgan
            Kaufman, 1991.

[Dur96]   S. Durand and F. Alexandre, TOM, A New Temporal Neural Net
            Architecture for Speech Signal Processing, in *Proceedings of ICASSP '96*,
            Vol. 6, pp 3550-3553.

[Elm90]   J. L. Elman, Finding Structure in Time, *Cognitive Science*, Vol. 14, pp
            179-211, 1990.

[Eul96a]   N.R. Euliano and J.C. Principe, Spatio-Temporal Self-Organizing Feature
            Maps, in *Proceedings of the ICNN '96*, Washington DC, June 1996, pp.
            1900-1905.

[Eul96b]   N.R. Euliano, J.C. Principe, P. Kulzer, A Self-Organizing Temporal
            Pattern Recognizer with Application to Robot Landmark Recognition,
            Accepted to the Sintra *Spatiotemporal Models in Biological and Artificial
            Systems Workshop*, November, 1996.

[Fah91]   S.E. Fahlman and C. Lebiere, The Cascade-Correlation Learning
            Architecture, Carnegie Mellon University, Computer Science Technical
            Report CMU-CS-90-100, 1991.

[Far95]   I. Farkas, *On Vector-coded Feature Mapping Using Self-Organizing
            Neural Maps*, Ph.D. Thesis, Slovak Technical University, Bratislava,
            1995.

[Fre91]   J.A. Freeman and D.M. Skapura, *Neural Networks: Algorithms,
            Applications, and Programming Techniques*, Addison-Wesley, Reading,
            MA, 1991.

[Fre92]   W.J. Freeman, Tutorial on Neurobiology: From Single Neurons to Brain Chaos, *International Journal of Bifurcation and Chaos*, Vol. 2, No.3, pp. 451-482, 1992.

[Fuk88]   K. Fukushima, Neocognitron: A Hierarchical Neural Network Capable of Visual Pattern Recognition, *Neural Networks*, Vol. 1, pp. 119-130, 1988.

[Gil91]   C. Giles, D. Chen, C. Miller, H. Chen, G. Sun, Y. Lee, Second Order Recurrent Networks for Grammatical Inference ", *Int. Joint Conf. On Neural Nets*, Vol. 2, pp. 273-281, Seattle, WA, 1991.

[Gop94a]   J. Goppert and W. Rosenstiel, Selective Attention and Self-Organizing Maps, in *Proceedings of Neural Networks and their Applications "*, IUSPIM, Marseille, France, 1994.

[Gop94b]   J. Goppert and W. Rosenstiel, Dynamic Extensions of Self-Organizing Maps, in *Proceedings of the International Conference on Artificial Neural Networks*, Sorrento, Springer, London 1994.

[Gop95]   J. Goppert and W. Rosenstiel, Neurons with Continuous Varying Activation in Self-Organizing Maps, in *From Natural to Artificial Neural Computation*, Lecture notes in Computer Science, Vol. 930, pp. 416-426, Springer-Verlag, 1995.

[Gro82]   S. Grossberg, Learning by Neural Networks. In S. Grossberg, editor, *Studies of Mind and Brain*, D. Reidel Publishing, Boston MA, pp. 65-156, 1982.

[Hay94]   S. Haykin, *Neural Networks: A Comprehensive Foundation*, Macmillan College Publishing Company, 1994.

[Hay96]   S. Haykin, *Adaptive Filter Theory*, 3rd edition, Prentice-Hall, pp.562-588, 1996.

[Haf90]   P. Haffner, M. Franzini, A. Waibel, Integrating Time Alignment and Neural Networks for High Performance Continuous Speech Recognition, Proceedings of ICASSP, Vol. 1, pp 425-428, IEEE, 1990.

[Hec86]   R. Hecht-Neilsen, Nearest matched filter classification of spatiotemporal patterns. Technical report, Hecht-Neilsen Neurocomputer Corporation, San Diego, CA June 1986.

[Hec90]   R. Hecht-Neilsen, *Neurocomputing*, Addison-Wesley Publishing Company, p. 168, 1990.

[Het93]   P. A. Hetherington and M. L. Shapiro, A Simple Network Model Simulates Hippocampal Place Fields: II. Computing Goal-Directed

Trajectories and Memory Fields, *Behavioral Neuroscience*, Vol. 107, No. 3, pp. 434-443, 1993.

[Jan97]   J.-S.R. Jang, C.-T. Sun, E. Mizutani, *Neuro-Fuzzy and Soft Computing: A Computational Approach to Learning and Machine Intelligence*, Prentice Hall, pp. 523-533, 1997.

[Jor86]   M.I. Jordan, Attractor Dynamics and Parallelism in a Connectionist Sequential Machine, in *Proceedings of the 9th Annual Conference of the Cognitive Science Society*, pp. 531-546, 1986.

[Kan90]   J. Kangas, Time-Delayed Self-Organizing Maps, in *Proceedings of the International Joint Conference on Neural Networks*, pp. 331-336, part 2 of 3, 1990.

[Kan91]   J.Kangas, Phoneme Recognition Using Time-Dependent Versions of Self-Organizing Maps, in *Proceedings of the International Conference on Acoustic and Speech Signal Processing*, Vol. 1, pp. 101-104, 1991.

[Kan92]   J. Kangas, Temporal Knowledge in Locations of Activations in a Self-Organizing Map, in *Artificial Neural Networks 2*, pp. 117-120, 1992.

[Kan94]   J. Kangas, On the Analysis of Pattern Sequences by Self-Organizing Maps, Unpublished Ph.D. Dissertation, Helsinki University of Technology, 1994.

[Kar94]   H. Kargupta and S. R. Ray, Temporal Sequence Processing Based on the Biological Reaction-Diffusion Process, *Proceedings of the IEEE ICNN '94*, Vol. 4, pp. 2315-2320, 1994.

[Koh82]   T. Kohonen, Self-Organized Formation of Topologically Correct Feature Maps, *Biological Cybernetics*, Volume 43, pp. 59-69, 1982.

[Koh91]   T. Kohonen, The Hypermap Architecture, in *Proceedings of the International Conference on Artificial Neural Networks*, pp. 1357-1360, 1991.

[Kre96a]   B. Krekelberg and J.G. Taylor, Nitric Oxide and the Development of Long-Range Horizontal Connectivity, *Neural Networks World*, Vol. 6, No. 2, pp. 185-189, 1996.

[Kre96b]   B. Krekelberg and J.G. Taylor, Nitric Oxide in Cortical Map Formation, *International Confernance on Artificial Neural Networks, 1996.*

[Kul96]   P. Kulzer, "NAVBOT – Autonomous robotic agent with neural network learning of autonomous mapping and navigation strategies", unpublished Master's Thesis from the University of Aveiro, Portugal, 1996.

[Mac62]   MacKay, Self-Organization in the Time Domain, *Self-Organizing Systems*, 1962.

[Mar90]   T.M. Martinetz and K.J. Schulten, Hierarchical neural net for learning control of a robot's arm and gripper, in the proceedings of *International Joint Conference on Neural Networks 90*, pp. 747-752, 1990.

[Mar93]   T.M. Martinetz, S.G. Berkovich, K.J. Schulten, "Neural-Gas" Network for Vector Quantization and its Application to Time-Series Prediction, *IEEE Transactions on Neural Networks*, Vol. 4, No. 4, July 1993, pp. 558-569.

[McA94]   J.D. McAuley, Time as Phase: A Dynamic Model of Time Perception, in *Proceedings of the 16th Annual Conference of the Cognitive Science Society*, Lawrence Erlbaum, pp 607-612, 1994.

[Miy93]   H. Miyamoto and K. Fukushima, Recognition of Spatio-Temporal Patterns by a Multi-Layered Neural Network Model, *Proceedings of the 1993 International Joint Conference on Neural Networks*, pp. 2267-2270, 1993.

[Moz92]   M.C. Mozer, Induction of Multiscale Temporal Structure, in *Advances in Neural Information Processing Systems IV*, pp. 275-282, published by Morgan Kaufmann, 1992.

[Moz94]   M. Mozer, Neural Net Architectures for Temporal Sequence Processing, in A.S. Weigend, N.A. Gershenfeld (eds) *Time Series Prediction: Forecasting the Future and Understanding the Past*, pp. 243-264, Addison Wesley Publishing Company, 1994.

[Mur89]   J. Murray, *Mathematical Biology*, Springer-Verlag, New York, 1989.

[Ore94]   R. C. O'Reilly and J. L. McClelland, Hippocampal Conjunctive Encoding, Storage, and Recall: Avoiding a Tradeoff, Parallel Distributed Processing and Cognitive Neuroscience Technical Report PDP.CNS.94.4, June 1994.

[Pea95]   B. Pearlmutter, Gradient Calculations for Dynamic Recurrent Neural Networks: A Survey. IEEE Transactions on Neural Networks, Vol. 6, No 3, pp. 1212-1228, 1995.

[Pol91]   J.B. Pollack, The Induction of Dynamical Recognizers, *Machine Learning*, Vol. 7, pp. 227-252, 1991.

[Pri94]   C. M. Privitera, P. Morasso, The Analysis of Continuous Temporal Sequences by a Map of Sequential Leaky Integrators, *Proceedings of ICNN 94*, pp. 3127-3130, 1994.

[Pri96]   C.M. Privitera and L. Shastri, A DSOM Hierarchical Model for Reflexive Processing: An Application to Visual Trajectory Classification,

International Computer Science Institute, Berkeley CA, Technical Report TR-96-011, June 1996.

[Rei91]   M. Reiss and J.G. Taylor, Storing Temporal Sequences, *Neural Networks*, Vol4, pp 773-787, 1991.

[Rol93]   E.T. Rolls and A. Treves, Neural Networks in the Brain Involved in Memory and Recall, in *Proceedings of the 1993 International Joint Conference on Neural Networks*, page 9-14, 1993.

[Rum86]   D.E. Rumelhart, G.E. Hinton, R.J. Williams, Learning Internal Representations by Error Propagation, in D.E. Rumelhart and J.L McClelland eds., *Parallel Distributed Processing: Explorations in the Microstructure of Cognition*, Vol. 1, Chapter 8, Cambridge, MA, MIT Press, 1986.

[Ruw93]   D. Ruwisch, M. Bode, H.-G. Purwins, Parallel Hardware Implementation of Kohonen's Algorithm with an Active Medium, *Neural Networks*, bol. 6, pp. 1147-1157, 1993.

[Ruw97]   D. Ruwisch, B. Dobrzewski, & M. Bode, Wave Propagation as a Neural Coupling Mechanism: Hardware for Self-Organizing Feature Maps and the Representation of Temporal Sequences, in IEEE Workshiop on Neural Networks for Signal Processing Proceedings, pp. 306-315, 1997.

[San97]   I.W. Sandberg & L. Xu, Uniform Approximation and Gamma Networks, *Neural Networks*, Vol. 10, pp. 781-784, 1997.

[Saw91]   H. Sawai, Frequency Shift Invariant Time-Delay Neural Networks for Robuts Continuous Speech Recognition, Proceedings of ICASSP, Vol. 1, IEEE, 1991.

[Sch90]   N. A. Schmajuk, Role of the Hippocampus in Temporal and Spatial Navigation: an Adaptive Neural Network, *Behavioral Brain Research*, Vol. 39, pp 205-229, Elsevier, 1990.

[Sch92a] J. Schmidhuber, Learning Unambiguous Reduced Sequence Descriptions, in *Advances in Neural Information Processing Systems IV*, pp. 291-298, published by Morgan Kaufmann, 1992.

[Sch92b] J. Schmidhuber, Learning to Control Fast-Weight Memories: An Alternative to Dynamic Recurrent Networks, *Neural Computation*, Vol. 4, No. 1, pp. 131-139, 1992.

[Sch92c]   J. Schmidhuber, A Fixed Size Storage O(n3) Time Complexity Learning
            Algorithm for Fully Recurrent Continually Running Networks. Neural
            Computation 4, pp. 243-248, 1992.

[Sha91]   P.E. Sharp, Computer Simulation of Hippocampal Place Cells,
            Psychobiology, Vol. 19, No. 2, pp 103-115, 1991.

[Sha93]   M. L. Shapiro and P. A. Hetherington, A Simple Network Model
            Simulates Hippocampal Place Fields: Parametric Analysis and
            Physiological Predictions, *Behavioral Neuroscience*, Vol. 107, No. 1, pp
            34-50, 1993.

[She94]   A. Shertinsky, R. W. Picard, M-Lattice: A Novel Non-Linear Dynamical
            System and Its Application to Halftoning, *Proceedings of IEEE ICASSP*,
            Vol. II, pp 565-568, 1994.

[Sta75]   J.C. Stanley and W.L Kilmer, A Wave Model of Temporal Sequence
            Learning, *International Journal of Man-Machine Studies*, Vol. 7, pp. 395-
            412, 1975.

[Sun92]   G.Z. Sun, H.H Chen, Y.C. Lee, Green's Function Method for Fast On-line
            Learning Algorithm of Recurrent Neural Networks. In NIPS 4, pp. 333-
            340, 1992.

[Tak81]   F. Takens, *Dynamical Systems and Turbulence*, in D.A. Rand and L.S.
            Rand (eds.), Vol. 898 of *Lecture Notes in Mathematics*, Springer Verlag,
            Berlin, 1981.

[Tan87]   D.W. Tank and J.J. Hopfield, "Neural Computation byConcentrating
            Information in Time", *Proceedings of the National Academy of Sciences,
            USA*, Vol. 84, pp 1896-1900, 1987.

[Tsu93]   M. Tsukada, A Theoretical Model of the Hippocampal-Cortical Memory
            System Motivated by Physiological Functions in the Hippocampus, in
            *Proeedings of the 1993 International Joint Conference on Neural
            Networks*, pp 1120-1123.

[Tur52]   A. Turing, The Chemical Basis of Morphogenesis, *Phil. Transactions of
            the Royal Society of London*, Ser. B, Vol. 237, pp. 37-72, 1952.

[Tys88]   J.J Tyson and J.P Keener, Singular Perturbation Theory of Traveling
            Waves in Excitable Media (A Review), *Physica D*, Vol. 32, pp 327-361,
            1988.

[Wai89a]   A. Waibel, Modular Construction of Time-Delay Neural Networks for
            Speech Recognition, *Neural Computation*, Vol. 1, pp 39-46, 1989.

[Wai89b]   A. Waibel, T. Hanazawa, G.E. Hinton, K. Shikano, K.J. Lang, Phoneme Recognition Using Time-Delay Neural Networks, 1989.

[Wan90]   D. Wang and M.A. Arbib, Complex Temporal Sequence Learning Based on Short-term Memory, in *Proceedings of the IEEE*, 1990.

[Wan93]   D. Wang and M.A. Arbib, Timing and Chunking in Processing Temporal Order, IEEE Transaction on Systems, Man, and Cybernetics, 1993.

[Wan95]   D. Wang, Temporal Pattern Encoding in Neural Networks, Wang, in *Handbook of Brain Theory and Neural Networks*, MIT Press, 1995.

[Wat90]   R. L. Watrous, Phoneme Discrimination Using Connectionist Networks, *Journal of the Acoustic Society of America*, Vol. 87, N 4. pp. 1753-1772, April 1990.

[Wat91]   R. L. Watrous, Context-Modulated Vowel Discrimination Using Connectionist Networks, *Computer Speech and Language*, Vol. 5, pp 341-362, 1991.

[Wei94]   A.S. Weigend, N.A. Gershenfeld (eds) *Time Series Prediction: Forecasting the Future and Understanding the Past*, Addison Wesley Publishing Company, 1994.

[Wil76]   D.J. Willshaw and C. von der Malsberg, How patterned Neural Connections can be Setup Up by Self-Organization, *Proceedings of the Royal Society of London, Series B*, Vol. 194, pp. 431-445, 1976.

[Wil89]   R.J. Williams and D. Zisper, A Learning Algorithm for Continually Running Fully Recurrent Neural Networks. *Neural Computation*, Vol. 1, pp 270-280, 1989.

[Wil90]   R.J. Williams and J. Peng, "An Efficient Gradient-Based Algorithm for On-Line Training of Recurrent Network Trajectories", *Neural Computation*, Vol. 2, pp. 490-501, 1990.

[Yam93]   K. Yamauchi, M. Fukuda, K. Fukushima, A Speech Recognition System Consisting of Auditory Feature Extracting Cells and Velocity-Controlled Delay-Lines Part II. Recognition Model, *Proceedings of the 1993 International Joint Conference on Neural Networks*, pp. 259-262, 1993.

[Zip89]   D. Zipser, A Subgrouping Strategy that Reduces Complexity and Speeds Up Learning in Recurrent Networks. Neural Computation 1, pp. 552-558, 1989.

[Zip90]   D. Zipser, Subgrouping Reduces Complexity and Speeds up Learning in Recurrent Networks. In NIPS 2, pp. 638-641, 1990.

BIOGRAPHICAL SKETCH

Neil R. Euliano II was born in Erie, Pennsylvania, April 2, 1964. Neil attended Lake Brantley High School in Orlando, and graduated as the academic salutatorian. Neil continued his education at the University of Florida receiving a bachelor's degree in computer engineering. In August 1986 he was recognized as the Four Year Scholar of his graduating class. In 1986 Neil became the first recipient of the Challenger Memorial Scholarship for graduate study in the field of electrical engineering. He graduated with a Masters of Engineering in electrical engineering in August 1988. His master's thesis was titled "The Implementation of General-Purpose Systolic Arrays for Digital Signal Processing and Linear Algebra."

Neil worked for AT&T Bell Laboratories in the Avionics Systems Division as a Systems Engineer from 1988 until 1992, when he returned to the University of Florida for his Ph.D. After graduation Neil will remain in Gainesville working as a consultant for two local startup companies. Neil has a wife, Tammy, and two children, Erin (age 2 ½) and Matthew (age 1 ½).

I certify that I have read this study and that in my opinion it conforms to acceptable standards of scholarly presentation and is fully adequate, in scope and quality, as a dissertation for the degree of Doctor of Philosophy.

Jose C. Principe, Chairman
Professor of Electrical and
Computer Engineering

I certify that I have read this study and that in my opinion it conforms to acceptable standards of scholarly presentation and is fully adequate, in scope and quality, as a dissertation for the degree of Doctor of Philosophy.

Fred J. Taylor
Professor of Electrical and
Computer Engineering

I certify that I have read this study and that in my opinion it conforms to acceptable standards of scholarly presentation and is fully adequate, in scope and quality, as a dissertation for the degree of Doctor of Philosophy.

Donald G. Childers
Professor of Electrical and
Computer Engineering

I certify that I have read this study and that in my opinion it conforms to acceptable standards of scholarly presentation and is fully adequate, in scope and quality, as a dissertation for the degree of Doctor of Philosophy.

William W. Edmonson
Assistant Professor of Electrical and
Computer Engineering

I certify that I have read this study and that in my opinion it conforms to acceptable standards of scholarly presentation and is fully adequate, in scope and quality, as a dissertation for the degree of Doctor of Philosophy.

Jon C. Allen
Professor of Entomology and
Nematology

This dissertation was submitted to the Graduate Faculty of the College of Engineering and to the Graduate School and was accepted as partial fulfillment of the requirements for the degree of Doctor of Philosophy.

August 1998

Winfred M. Phillips
Dean, College of Engineering

Karen A. Holbrook
Dean, Graduate School

www.ingramcontent.com/pod-product-compliance
Lightning Source LLC
Chambersburg PA
CBHW080414060326
40689CB00019B/4234